Securing Next-Generation Connected Healthcare Systems

Securing Next-Generation Connected Healthcare Systems

Artificial Intelligence Technologies

Edited by

DEEPAK GUPTA
Department of Computer Science and Engineering,
Maharaja Agrasen Institute of Technology, Delhi, India

ABOUL ELLA HASSANIEN
College of Business Administration, Kuwait University,
Kuwait

ELSEVIER

ACADEMIC PRESS
An imprint of Elsevier

ISBN: 978-0-443-13951-2

For Information on all Academic Press publications
visit our website at https://www.elsevier.com/books-and-journals

Publisher: Mara Conner
Acquisitions Editor: Carrie Bolger
Editorial Project Manager: Isabella Silva
Production Project Manager: Sujithkumar Chandran
Cover Designer: Mark Rogers

Typeset by MPS Limited, Chennai, India

Working together to grow libraries in developing countries
www.elsevier.com • www.bookaid.org

Contents

List of contributors

A. Aleem
Department of CSE, UIE, Chandigarh University, Mohali, Punjab, India

A.N. Arun
Department of Computer Science & Engineering, Sri Venkateswara Institute of Science and Technology, Tiruvallur, Tamil Nadu, India

V. Suresh Babu
Department of Electronics and Communication Engineering, Kalasalingam Academy of Research and Education, Krishnankoil, Tamil Nadu, India

Susama Bagchi
Chitkara University Institute of Engineering and Technology, Chitkara University, Punjab, India

Ankit Bansal
Chitkara University Institute of Engineering and Technology, Chitkara University, Punjab, India

Priyanka Bhutani
University School of Information and Communication Technology, GGSIPU, Delhi, India

Sunil Kumar Chawla
Chitkara University Institute of Engineering and Technology, Chitkara University, Punjab, India

Pragyan Das
Kalinga Institute of Industrial Technology, Deemed to be University, Bhubaneswar, Odisha, India

Sanjoy Kumar Debnath
Chitkara University Institute of Engineering and Technology, Chitkara University, Punjab, India

Priyanka Dhaka
University School of Information and Communication Technology and Maharaja Surajmal Institute, GGSIPU, Delhi, India

Ankit Garg
AIT-CSE, Chandigarh University, Mohali, Punjab, India; University Center for Research and Development, Chandigarh University, Mohali, Punjab, India

Ishika Gupta
Kalinga Institute of Industrial Technology, Deemed to be University, Bhubaneswar, Odisha, India

A.M. Gurusigaamani
Department of Computer Science Engineering, Kalasalingam Academy of Research and Education, Krishnankoil, Tamil Nadu, India

Danish Jamil
Department of Information Technology, Malaysia University of Science and Technology, Petaling Jaya, Malaysia

T. Jayasankar
Department of Electronics and Communication Engineering, Anna University, Trichy, Tamil Nadu, India

Tejinder Kaur
Chitkara University Institute of Engineering and Technology, Chitkara University, Punjab, India

Vikash Kumar
Kalinga Institute of Industrial Technology, Deemed to be University, Bhubaneswar, Odisha, India

P. Maheswaravenkatesh
Department of Electronics and Communication Engineering, Anna University, Trichy, Tamil Nadu, India

R. Manikandan
School of Computing, SASTRA Deemed University, Thanjavur, Tamil Nadu, India

Sushruta Mishra
Kalinga Institute of Industrial Technology, Deemed to be University, Bhubaneswar, Odisha, India

A. Mohit
AIT-CSE, Chandigarh University, Mohali, Punjab, India; University Centre for Research & Development (UCRD), Chandigarh University, Mohali, Punjab, India

Suvra Mukherjee
Kalinga Institute of Industrial Technology, Deemed to be University, Bhubaneswar, Odisha, India

Ayush Pal
Kalinga Institute of Industrial Technology, Deemed to be University, Bhubaneswar, Odisha, India

Sellappan Palaniappan
Department of Information Technology, Malaysia University of Science and Technology, Petaling Jaya, Malaysia

D. Prabakar
Department of Data Science and Business Systems, School of Computing, Faculty of Engineering and Technology, SRM Institute of Science and Technology, Chengalpattu, Tamil Nadu, India

Shamimul Qamar
Computer Science & Engineering Department, College of Sciences & Arts, Dhahran Al Janoub Campus, King Khalid University, Abha, Kingdom of Saudi Arabia

Jenyfal Sampson
Department of Electronics and Communication Engineering, Kalasalingam Academy of Research and Education, Krishnankoil, Tamil Nadu, India

Ruchi Sehrawat
University School of Information and Communication Technology, GGSIPU, Delhi, India

Anuj Kumar Singh
Amity University, Gwalior, Madhya Pradesh, India

Greeshmitha Vavilapalli
Kalinga Institute of Industrial Technology, Deemed to be University, Bhubaneswar, Odisha, India

S.P. Velmurugan
Department of Electronics and Communication Engineering, Kalasalingam Academy of Research and Education, Krishnankoil, Tamil Nadu, India

P. Vigneshwaran
Department of Electronics and Communication Engineering, Kalasalingam Academy of Research and Education, Krishnankoil, Tamil Nadu, India

CHAPTER 1

Authentication protocols for securing IoMT: current state and technological advancements

Anuj Kumar Singh[1] and Ankit Garg[2,3]
[1]Amity University, Gwalior, Madhya Pradesh, India
[2]AIT-CSE, Chandigarh University, Mohali, Punjab, India
[3]University Center for Research and Development, Chandigarh University, Mohali, Punjab, India

1.1 Introduction

The Internet of Medical Things, or IoMT for short, is a network of hardware infrastructures, application programs, and clinical devices connected to the Internet that is used to link together health-care information technology system. IoMT, also known as IoT in the health-care sector, makes it feasible for wireless and distant equipment and device to communicate securely via the Internet to enable instantaneous and adaptable analysis of medical data. Integrated medical equipment assists health-care professionals in providing quicker and more effective care. The activities like robotic surgery and glucose monitoring are now possible owing to the advances in IoMT. Superior treatments and cost reductions are two primary advantages of IoMT. Improved and regular monitoring of patients without the need for visits to medical centers and faster and more accurate diagnosis are possible due to IoMT technology. IoMT has a wide range of effects in the medical field. The use of IoMT in the home, on the body, in the hospital, and in the community most clearly demonstrates these changes.

- IoMT in the home—The capability to communicate health-related information from patient's house to another entity, which might be a hospital or patient's primary care physician, is provided by in-home IoMT.
- IoMT on the body—Wearable medical equipment that is linked to remote tracking or monitoring systems is used in on the body IoMT. On the body, IoMT, as compared to in the home IoMT, is frequently utilized while people go outside the boundary of home for their daily activities.

Securing Next-Generation Connected Healthcare Systems
DOI: https://doi.org/10.1016/B978-0-443-13951-2.00004-0

- IoMT in the hospital—Hospitals need to continually regulate the accessibility and efficacy of their clinical resources, as well as the flow of both medical staff and patients through their establishment. To provide administrators with an accurate picture of what is happening, health-care experts track all of these interactions using IoMT sensors and other tracking devices.
- IoMT in the community—The use of IoMT devices over a larger town or region is known as community IoMT. Technology that enables remote services is also a part of community IoMT, in addition to mobile and emergency care. To facilitate the transportation of medical supplies and equipment, suppliers may also deploy IoMT devices in logistics.

Safeguarding sensitive and personal data is one of the main security issues for wearable and implantable devices. Health, location, and biometric data are just a few of the types of information that these devices gather and store. To avoid potential misuse or exploitation, it is crucial to ensure that these data are kept secret and shielded from unwanted access. To protect this sensitive information and stop unauthorized manipulation or interception, manufacturers must employ effective encryption and authentication procedures. In addition, wearable and implanted tech frequently connects to other hardware or networks, including cloud-computing services, cell phones, and tablets. Additional points of vulnerability are introduced by these interconnections, which bad actors could use. The most important steps to lessen risks are to ensure secure communication routes and to update software often to fix flaws. Owing to the use of wearable and implantable devices that are low on computing capability and transmission of data over open communication channel, there are various security concerns with IoMT [1].

1.1.1 Applications of Internet of Medical Things

IoMT that connects medical equipment, software, and apps to the Internet and other linked technologies has completely transformed the health-care sector. Due to this convergence a wide range of cutting-edge applications have been developed that increase patient care, simplify medical procedures, and boost the effectiveness of health-care systems. Applications for IoMT are numerous and extensive, covering many facets of healthcare.

- *Remote patient monitoring:* It is one of IoMT's main applications. Health-care providers can monitor patients' vital signs, activity levels, and medication adherence from a distance with the aid of wearable technology, sensors, and connected health platforms. With the use of this real-time data, health-care professionals are better equipped to recognize early warning signals, take swift action, and tailor patient care to their unique needs.

- *Telemedicine:* A crucial application that permits online discussions between patients and medical professionals is called telemedicine. Especially in rural or underserved locations, IoMT enables secure and effective medical data transfer, enabling remote diagnosis, and minimizing the need for in-person consultations. It also makes health-care services more accessible while saving time and money for both individuals and health-care providers.

- *Hospital efficacy and patient's safety:* Additionally, IoMT is essential for enhancing patient safety and hospital productivity. A hospital's network can be linked with smart medical devices, including infusion pumps, to enable real-time monitoring of equipment status and upkeep requirements. Additionally, IoMT can aid with asset tracking, ensuring that medical supplies are available when needed and lowering the possibility of medical errors.

- *Smart pharmaceutical industry:* IoMT also has advantages for the pharmaceutical industry. Smart pill bottles and medication adherence tools can help patients remember to take their pills on time and can give medical professionals useful information on patient compliance. This facilitates improved chronic condition management and guards against potential health issues.

1.1.2 Security in Internet of Medical Things

Medical devices are becoming more commonplace in IoMT, which has several advantages but at the same time raises severe privacy and security issues. Medical information of the patient that is sensitive and also frequently life-critical is gathered, processed, and used to influence crucial decisions by health-care organizations. By exploiting the security vulnerabilities found in these IoMT devices, cybercriminals have a chance to not only get access to the health-care system, but also to sensitive patient and health data. Attacks on these interconnected IoMT devices have the potential to endanger the patients' lives and cause serious physical harm

[2]. Concerns related to IoMT can be categorized into four major groups, one of which is brought up by the general public and encompasses difficulties with the issues of security, privacy, accuracy, and trust [3]. These concerns have been highlighted in Fig. 1.1.

1.1.2.1 Security
Because of the inadequacy of protective security measures that the majority of IoMT devices can suffer from due to the design or owing to inadequate secure authentication mechanisms that may be readily circumvented by a competent attacker, an attacker can eavesdrop and capture both inbound and outbound data and information across IoMT. Because of the inability to identify and counteract such assaults, it is also feasible to get unauthorized access without being observed. This could lead to higher privileges to the attacker, the injection of harmful code, or the malware penetration of devices. To be able to guarantee and preserve the safety of the medical-cyber physical system, as well as medical devices and networks, manufacturers of health-care products must prioritize security as a fundamental task. To minimize the primary IoMT security issues, guarding against passive and active attacks is essential.

1.1.2.2 Privacy
Considering the possibility of collecting and disclosing patient identities along with sensitive and private data, passive attacks like traffic monitoring raise privacy concerns. The capability of an adversary to identify a patient's private medical data and illnesses poses a grave threat to the patient's life, making this a very significant threat. Theft of identities is yet another reason why patients' privacy is violated when hospitals are attacked. In almost every one of these real-life attacks, sensitive or confidential data were leaked or otherwise revealed, which caused a breach of patients' privacy.

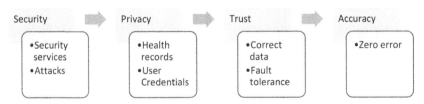

Figure 1.1 Internet of Medical Things concerns.

1.1.2.3 Trust

The privacy of patients being compromised results in significant trust challenges. Patients have become skeptical of the notion that technology will eventually take the place of humans in positions such as physicians, nurses, and office staff. People are consequently more anxious about having a clinical robot, device, or even gadget keeping track of their health concerns.

1.1.2.4 Accuracy

Lack of accuracy and diagnosis in medical robots due to unintentional errors has become a serious problem. Numerous incidences have been documented demonstrating the lack of precision and accuracy in the procedures being directed by medical robots, coupled with patient misdiagnosis and incorrect prescriptions.

1.1.3 Security requirements for Internet of Medical Things

National Institute of Standards and Technology (NIST) specifies confidentiality, availability, and integrity as the fundamental security requirement for any system. However, in practical situations, more security attributes and functions need to be satisfied. Si–Ahmed et al. [4] highlighted that in addition to the basic security attributes specified by NIST, IoMT system should also satisfy the properties of forward security, data freshness, scalability, authentication, nonrepudiation, authorization, and auditability. The complete set of security attributes required by IoMT system has been shown in Fig. 1.2.

- *Confidentiality*—The data originating from IoMT devices must not be disclosed to unauthorized party.
- *Integrity*—The data originating from IoMT devices must not be altered during transit.
- *Availability*—The medical data from the devices can be obtained anytime by the authorized users.
- *Forward security*—Regardless of a long-term session key being compromised, an attacker cannot access the previous messages.
- *Authorization*—A privilege given to an entity allowing them to access the IoMT device data [5].
- *Authentication*—Before accessing the IoMT device and data, the legitimate user must be authenticated to confirm the identity [6].

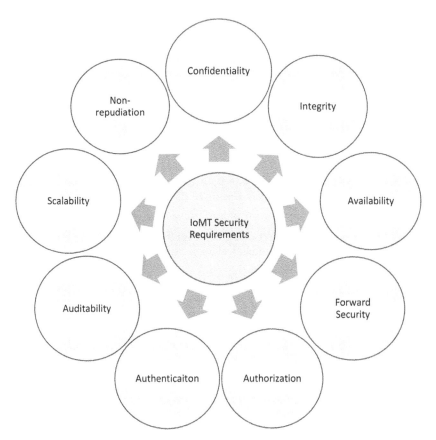

Figure 1.2 Security requirements of Internet of Medical Things (IoMT) [4].

- *Auditability*—IoMT systems must be competent to review and analyze security-related records and operations for the purpose of evaluating the efficacy of system controls, ensure adhering to established security procedures and policies, find potential security breaches, and suggest any changes that are necessary for defensive measures.
- *Scalability*—The IoMT system must be capable of accommodating additional devices in the system.
- *Nonrepudiation*—After transmitting/receiving the data, IoMT device as well as the user must not be able to deny the same.

1.1.4 Attacks on Internet of Medical Things

Papaioannou et al. [7] performed an analysis over the attacks attempted on IoMT system and identified that attacks can be grouped as attacks on data confidentiality, attacks on integrity, attacks on authentication, attacks on authorization, and attacks on availability. This classification of attacks has been elucidated in Table 1.1.

As the IoMT ecosystem expands to include a variety of connected devices and apps, it opens itself up to a wide range of potential threats. Authentication is a crucial security measure that defends against unauthorized access to and manipulation with private medical data. The security of authentication systems in IoMT is seriously threatened by numerous assaults, including credential stuffing, man-in-the-middle attacks, and brute force attacks. Strong authentication protocols, multifactor authentication (MFA) systems, and encryption approaches must be adopted to address these issues to strengthen the security of IoMT networks and thwart future breaches. A resilient and secure IoMT infrastructure must also be maintained, and this includes ongoing monitoring, regular security audits, and fast updates to address emerging risks. The IoMT can keep revolutionizing health-care delivery while preserving patient privacy and data integrity by placing a high priority on the protection of authentication mechanisms.

1.2 Authentication in Internet of Medical Things

The most significant security attribute in the IoMT environment is authentication because if appropriate authentication is applied in a secure and efficient manner, it can prevent many attacks. It is difficult to come up with general authentication methods for different IoMT system nodes due to the scattered and complicated nature of IoMT systems. In the IoMT, authentication solutions are essential for preserving the confidentiality, security, and integrity of sensitive patient data and guaranteeing the consistent operation of connected medical devices. IoMT not only continues to transform healthcare by providing seamless data interchange and real-time monitoring, but it also poses special issues that call for strong authentication procedures.

The three layers of device, user, and network are where IoMT authentication is generally taken into account. Whether a centralized or distributed structure is required for authentication in IoMT applications, a centralized server needs to exist for centralized authentication for the purpose of recognizing and authenticating system entities. The process of authentication is carried out by a number of scattered nodes in the distributed design, in contrast. A flat or multilevel design can be employed for both centralized and distributed systems [8].

Table 1.1 Classification of attacks in Internet of Medical Things (IoMT) [7].

Attack category	Specific attack	Elucidation
Attacks on confiden tiality	Eavesdropping	The attacker listens secretly to the communication media for obtaining secret information.
	Traffic analysis	The patterns in the communication are used by the attacker to infer information about the confidential data.
	Impersonation	To gain access to the IoMT data, an attacker impersonates a legitimate entity.
Attacks on integrity	Man-in-the-middle	Attacker intervenes in the conversation between the IoMT device and the user and possesses the capability to change the data.
	Physical	An adversary with physical access to an IoMT device might modify its structure and alter its functions.
	Malware	By running malicious code on an IoMT device, an attacker could damage it.
Attacks on authenti cation	Forgery	For the purpose of misleading other entities the attacker uses this technique to establish a false identity and transfer fraudulent information.
	Sybil	Multiple fake identities are used by the attacker to access a legitimate IoMT device.
	Device cloning	From a legitimate device the attacker gathers secret information that can be used to make a significant number of clones.
	Masquerading	To access IoMT data an attacker impersonates an authorized user.
Attacks on autho rization	Social engineering	For the purpose of gaining access to users' medical device, an adversary may deceive the IoMT edge network and pose as legitimate.
	Malware	The linked IoMT devices could be compromised by the attacks by infecting them.
Attacks on availa bility	Denial of service	The IoMT system is made unavailable by sending the fake requests by the attacker.

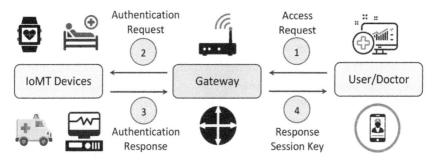

Figure 1.3 Authentication model for Internet of Medical Things (IoMT).

1.2.1 Authentication model for Internet of Medical Things

According to Ref. [9], the most suitable authentication model for IoMT environment has been demonstrated in Fig. 1.3. It depicts various components involved in ensuring secure and authorized access to medical devices and data within the IoMT ecosystem. Four messages are sent sequentially across the IoMT network to enable remote legitimate users for accessing the IoMT device. The remote user submits a request for accessing to the IoMT node to the gateway node in the initial message. The gateway node approaches the IoMT device for services in authenticating both itself and the remote user in the second message. The IoMT device informs the gateway with the response to this authentication request. In this method the IoMT node also authenticates the gateway. Finally, the gateway node gives the session key to the legitimate user. Let us assume that the mutual authentication process is a success. In that situation, there is no involvement from the gateway, and the data transmission or node access occurs directly between the remote user and the IoMT node.

The two primary elements of communication in cloud-based IoMT are outlined as follows: (1) only legitimate users have permission to access IoMT devices and acquire data from them and (2) the user connection and data communication need to be made by an authenticated IoMT device. The three security concerns that need to be resolved for secure communication among the remote user and the IoMT device are identity, privacy, and trust. Therefore establishing an effective mutual authentication system is crucial for this type of communication.

A brief description of different components used in the authentication of IoMT is presented next.

- *IoMT devices*: represent a range of medical equipment, including implantable devices, wearable health monitors, sensors, and systems, for remote patient monitoring. These gadgets gather and provide information about your health.

- *IoMT gateway*: serves as a bridge between IoMT devices and the main system. The gateway is in charge of gathering the data, encrypting it, and sending it to the right servers.
- *Central server*: a central system that receives, analyses, and saves medical data from IoMT devices. Additionally, it manages user credentials and handles requests for authentication.
- *User interface (UI)*: represents the UI that authorized users utilize to communicate with the main system, retrieve medical data, and manage IoMT devices.
- *Authentication server*: users' requests for authentication are handled by this server. It validates the user's credentials and authorizes access in accordance with the authentication procedure.
- *Authentication protocols*: depict several authentication technologies, such as OAuth, OpenID Connect, or SAML, which are used to securely identify and approve users.
- *Biometric authentication*: a second layer of protection that can be applied to identify users based on their distinctive physiological or behavioral traits, such as fingerprints, faces, or voices.
- *Two-factor authentication*: shows how to employ a secondary authentication technique, such as a one-time password or a token delivered to the user's mobile device, to increase security.
- *Secure communication*: represents encrypted communication routes (such as Transport Layer Security [TLS]/secure sockets layer [SSL]) that are used to provide safe data transmission between IoMT devices, gateways, and the main server.
- *Access control*: describes the system in place to limit user access to particular medical data depending on the user's roles, permissions, and level of trust.
- *Database*: exemplifies how the central system stores user login information and access rights.

1.2.2 Authentication techniques in Internet of Medical Things

According to the direction of the verifying entities, authentication in IoMT systems can be classified. The IoMT system supports one-, two-, and three-way authentication options. In IoMT frameworks, authentication can rely on a fundamental, certificate-based, key-based, or

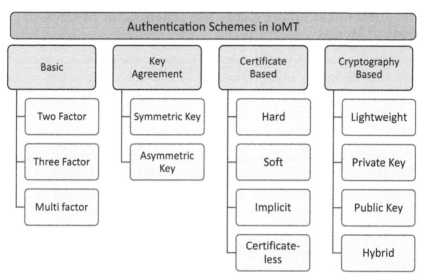

Figure 1.4 Classification of authentication methods in Internet of Medical Things (IoMT).

cryptography-based technique [10]. Recently, hybrid techniques have been employed by researchers to boost system security and performance. A categorization of authentication methods is shown in Fig. 1.4. In basic authentication techniques the characteristics used to identify a party are the same as those are used to authenticate it. Authentication in IoMT can also be achieved through creating a shared key that can be utilized by the communicating parties to interact securely. The application of certificates to identify legitimate entities can be an element of IoMT authentication systems. Therefore to recognize entities, methods of authentication may demand a hard certificate, implicit certificate, soft certificate, or no certificate at all. Authentication requires the use of cryptography, and different modern cryptography techniques present an excellent opportunity to enhance IoMT security as well as authentication [11].

Owing to the less computational capability of IoMT with respect to the processing power as well as memory, there is a need of authentication schemes that put minimum computational and communication overhead on IoMT system. Moreover, lightweight MFA schemes and protocols for IoMT utilizing the benefits of both symmetric key cryptosystem and public key cryptosystem should be considered while securing IoMT system [12].

1.3 Authentication protocols and schemes for Internet of Medical Things

Since the inception of IoMT, many authentication schemes and protocols have been devised by many researchers continuously. This section of the chapter provides an overview of the verity of authentication schemes and protocols that have been designed for IoMT environment.

1.3.1 Full-fledged authentication in Internet of Medical Things

These are comprehensive and strong security protocols intended to confirm the legitimacy of users trying to access a digital platform or service. These systems have multiple layers of authentication, generally combining the user's knowledge (such as a password), possession (such as a smartphone or hardware token), and intrinsic data (such as a fingerprint or other biometric information). Full-fledged authentication systems significantly improve security and defend against numerous threats, such as phishing, brute force attacks, and password leaks, by incorporating multiple aspects. By balancing security measures with ease of use, these advanced solutions not only guarantee the privacy and integrity of user data but also provide a seamless and user-friendly experience. In today's digital world, using full-fledged authentication techniques is crucial. Full-fledged authentication techniques, which protect patient privacy and stop unwanted access to critical medical information, include MFA, biometric verification, and cutting-edge encryption technologies. Health-care providers can confidently embrace the advantages of IoMT, promoting seamless data exchange, enabling remote patient monitoring, and enabling timely interventions, all while maintaining a high level of security and trust in this networked health-care environment, by implementing full-fledged authentication.

The two kinds of full-fledged authentication protocols that can be utilized for IoMT are TLS/SSL-based and public key infrastructure (PKI)-based. The first kind of cryptographic measures that offer secure network communication are TLS and SSL. To safeguard data while these are being transmitted, they create a secure channel between IoMT devices and servers. The second type of cryptographic method includes digital certificates and public−private key pairs that are managed using a set of infrastructure (PKI) policies, procedures, and technologies known as PKI. For secure authentication and communication in IoMT, PKI is essential.

1.3.2 Lightweight authentication in Internet of Medical Things

A protocol with a lesser and leaner payload when being utilized and communicated over a network connection is referred to as a lightweight protocol. Compared to other communication protocols used on a local or wide area network, it is simpler, faster, and easier to manage. Lightweight schemes and protocols are the ones that are employed for network communication with the least amount of computational and communication overhead. Despite the fact that lightweight protocols frequently offer the same or better services as their heavier counterparts, they generally leave a smaller environmental imprint overall. Code for lightweight protocols performs better than code for traditional protocols. They frequently use a data compression approach, are smaller overall, omit less important data, and have a less noticeable impact on network traffic. Some well-known examples of lightweight protocols include slim call control protocol, lightweight extensible authentication protocol, and lightweight directory access protocols.

The efficiency and simplicity of lightweight authentication methods are key design considerations. While lowering computational cost and memory needs, they seek to offer a high level of security. The main objective is to make sure that the authentication process does not put an undue burden on the target devices' finite resources. Therefore these techniques are especially attractive for resource-constrained devices, which are more common in contemporary computing environments. For achieving authentication for less computationally capable devices and environment, there is always a need for developing a mutual authentication scheme that consumes less computational time and bandwidth. Therefore developing a lightweight cryptographic authentication scheme is highly significant for IoMT environment. The lower computing complexity of lightweight authentication systems is one of their main advantages. Traditional authentication systems like RSA and ECC (elliptic-curve cryptography) demand computationally costly, large amounts of computing power, and memory. Lightweight methods, in contrast, use less resource-intensive and quicker cryptographic primitives such hash functions, symmetric ciphers, and lightweight cryptographic algorithms. This makes it possible to use simple authentication methods on hardware with low processing power without sacrificing security.

Lightweight authentication methods also have a low communication overhead as a core feature. Keeping the data transmitted during the authentication procedure to a minimum is crucial in situations when devices use low-bandwidth or unstable networks. Compact representations of authentication tokens are frequently used in lightweight methods to reduce the amount of data transmitted and the possibility of data transmission mistakes. They are therefore appropriate for use in networks with limited resources, where bandwidth is scarce and network connectivity is patchy. Taxonomy of lightweight authentication schemes highlighting the advantages and disadvantages has been presented in (Table 1.2).

Table 1.2 Taxonomy of lightweight authentication schemes.

Scheme name	Description	Advantages	Disadvantages
HMAC	Hash-based message authentication code	Simple and efficient	Requires a shared secret
OAuth	Open standard for token-based authentication	Allows third-party authorization	Complex for some imple mentations
JWT	JSON web tokens	Stateless and easy to implement	Lack of built-in session management
SRP	Secure Remote Password protocol	Strong against password guessing attacks	Heavier computational overhead
TOTP	Time-based one-time password	No network connectivity required	Synchronization issues with time drift
U2F	Universal 2nd factor	Phishing-resistant	Limited support in some applications
WebAuthn	Web authentication API	High security and user privacy	Requires modern browser support
Biometric Auth	Authentication based on biometric data	Convenient and user-friendly	Potential privacy and security concerns
Zero-Knowle-dge Proofs	Verifiable authentication without sharing passwords	High security with minimal data exposure	Complex to implement in some scenarios
Certificate-based	Authentication using digital certificates	Strong security and mutual authentication	Certificate management can be challenging

According to the authentication model shown in Fig. 1.3, the messages must be sent so that they consume minimum bits and they must not reveal any confidential information from these. Additionally, the security-related computations must take minimum time for creating the components to be sent within the messages [13,14]. It means there is a clear trade-off between the security functions of the security protocols and its efficiency.

1.3.3 Existing authentication schemes for Internet of Medical Things

Three communicating parties are used in the authentication and key establishment protocol created by Yeh et al. [15]. These are the server, the gateway that is also designated as the local processing unit, and the sensor node. The connection of medical experts to the server is not mentioned. As a result, it is implicitly expected that doctors will examine patient data while they are offline, meaning that there is no way to keep track of a patient in real time.

An authentication and key-generation protocol among a user, a gateway, and a sensor node was created by Gope and Hwang [16]. Anonymity is the main objective of this scheme. They refer to a gateway as a "server-class" device that links users to various sensor clusters. A sink node known as the cluster head serves as a conduit for communication between sensors in a cluster and an outside party.

Zaho [17] suggested an efficient ECC-based identification-based anonymous authentication mechanism for IoT-enabled healthcare. However, it was discovered that this technique is not safe against known session-specific temporary information attacks and cannot synchronize clocks. To protect medical sensor networks against known assaults, Kumar et al. [18] developed a two-factor technique. However, He et al. [19] showed that this mechanism is susceptible to insider attacks.

The gateway node of the patient's IoMT, the medical professional user, and a single sensor implanted within IoMT are the three communication participants for which Li et al.'s [20] authentication protocol is intended. The specification aims to safeguard user and sensor anonymity from the intruding party. The technique makes use of hashing and symmetric encryption/decryption, where a single key can be shared across a number of users. This use of keys implies that communication with all users or sensor nodes is compromised when one key is compromised. A doctor caring for N patients will need to register N times, which is not practicable. In addition, users must register themselves at the patient's WBAN gateway node.

For the purpose of anonymous key agreement and wireless channel data authentication, Rehman et al. [21] suggested a simple three-level cryptographic method. The suggested authentication strategy demonstrates how well it can defend against several known cyberattacks, including the base station compromise assault and the impersonation attack on the sensors. With the use of the AVISPA (Automated Validation of Internet Security Protocol and Applications) tool and BAN Logic, the system was formally validated.

In another research effort, Mucchi et al. [22] recommended an innovative modulation method using a thermal noise loop for safeguarding wireless communication at the physical layer. The proposed approach demonstrated resistance to the denial-of-service attacks and produced beneficial outcomes in a multiuser setting. For protected wireless communication, Soderi et al. [23] have presented a unique physical layer watermarking-based security system in conjunction with a jamming receiver. The outcomes showed that the protocol was energy-efficient and full-rate.

Ibrahim et al. [24] suggested a lightweight authentication mechanism that successfully meets the objectives and security requirements of an authentication service, as shown by BAN logic and informal security analysis. Throughput and end-to-end delay are just two examples of the network factors that the simulation results demonstrate their impact on. Several hundred bits must be stored by the network's nodes. Nodes are particularly efficient at computing because they only need to do a small number of hash invocations. The suggested protocol's one-round communication cost is a few hundred bits. The energy used by the nodes is likewise low because of the cheap cost of computation.

For IoMT WBANs, Xu et al. [25] presented a safe yet simple authentication mechanism. This method eliminates the need for asymmetric encryption and ensures forward secrecy. We examine informal security and verify the security of our scheme using the automatic security verification program ProVerif. Theoretical analysis and actual data show that our system has a lower security risk than lightweight schemes while dramatically reducing computing costs when compared to those using asymmetric encryption.

1.3.4 Security properties of authentication schemes for Internet of Medical Things

To protect against potential cyber risks and data breaches, strong authentication techniques are essential as IoMT technologies are increasingly being adopted in the health-care sector. The security and privacy of patients could be jeopardized by numerous attacks such as impersonation, replay, and man-in-the-middle assaults. These authentication techniques should exhibit resistance against these types of attacks. They should also be scalable, simple to use, and capable of handling the wide variety of networked devices seen in health-care contexts. The advancement of connected healthcare will be fueled by ensuring that these security features are met while keeping the highest standards of data protection and confidentiality [26]. This will provide health-care professionals, patients, and stakeholders the confidence to embrace IoMT advancements.

Table 1.3 compares the security features of various IoMT authentication techniques. Please keep in consideration that the security attributes listed in this table are only a general representation, depending on the specific implementation. Remember that adequate installation, configuration, and ongoing monitoring are necessary for these authentication methods' security properties to be effective in keeping up with the changing threat landscape.

1.3.5 Taxonomy of attacks on authentication schemes for Internet of Medical Things

To make sure that only authorized staff members may access patient data, manage medical equipment, or make crucial choices, IoMT devices significantly rely on authentication systems [27]. Cybercriminals break into the system, steal patient data, or modify medical devices for their own gain by taking advantage of weaknesses in authentication mechanisms, such as weak passwords, insufficient encryption, or unauthorized access attempts. Table 1.4 presents a taxonomy of different attacks on the authentication mechanism in IoMT environment. To protect the integrity and confidentiality of IoMT systems and preserve the public's confidence in connected health-care technologies, health-care providers and manufacturers must implement strong, MFA methods, continuous monitoring, and regular security updates [28].

To prevent these assaults, IoMT systems must have a multilayered security strategy that includes robust encryption, MFA, continuous monitoring, and frequent security updates. This will secure patient data and guarantee the reliability of the devices and networks.

Table 1.3 Security attributes, advantages, and limitations of different kinds of authentication schemes.

Authentication scheme	Security attributes	Advantages	Disadvantages
Username/ password	• Simple and familiar • Vulnerable to brute force attacks • Risk of password leakage	• Easy to implement • Low cost	• Prone to password reuse • Susceptible to social engineering attacks
Biometric	• Nontransferable and unique • Reduces password-related risks • Difficult to replicate • Multifactor authentication potential	• Difficult to forge • Enhanced user experience • Higher security level	• Biometric data can be compromised • Expensive hardware and software • False acceptance and rejection rates
RFID/NFC	• Contactless and convenient • Difficult to counterfeit • Enables seamless user experience • Can be used in conjunction with other methods	• Fast authentication process • No need for physical contact • Can be integrated into wearables	• Vulnerable to relay attacks • Limited range of communication • Limited data storage capacity
OTP	• Dynamic and timebound • Adds a layer of security to passwords • Reduces vulnerability to credential leaks	• Reduces risks of replay attacks • Works without Internet connectivity • Suitable for multifactor authentication	• Extra step in authentication process • Potential delivery delays of OTP • Possibility of interception

PKI	• Strong encryption and authentication • Utilizes digital certificates for identity • Provides secure key exchange • Robust protection against man-in-the-middle	• Highly secure • Suitable for large-scale systems • Difficult to forge	• Complex infrastructure setup • Certificate management overhead • Dependency on certificate authorities
OAuth	• Authorization framework for APIs • Token-based authentication • Supports delegation of access • Revocation of tokens possible	• Allows third-party access without sharing credentials • Widely adopted in web and mobile applications • Provides secure access to resources	• Can be complex to implement • Requires careful scope and permissions setup • Risk of token leakage or theft

OTP, One-time password; *PKI*, public key infrastructure.

Table 1.4 Taxonomy of attacks on authentication in Internet of Medical Things (IoMT).

Attack type	Description	Impact
Password guessing	Repeatedly trying different passwords to gain access.	Unauthorized access
Brute force	Exhaustively trying all possible combinations of inputs.	Unauthorized access
Dictionary attack	Using a pregenerated list of common passwords.	Unauthorized access
Phishing	Deceiving users into revealing their credentials.	Unauthorized access
MITM	Intercepting communication to capture credentials.	Unauthorized access, data manipulation
Replay attack	Capturing and reusing authentication data.	Unauthorized access, data manipulation
Credential theft	Stealing user credentials through various means.	Unauthorized access, data breach
Insider threat	Malicious actions by authorized personnel.	Unauthorized access, data breach
Session hijacking	Taking control of an active user session.	Unauthorized access, data manipulation
Device spoofing	Impersonating a legitimate IoMT device.	Unauthorized access, data manipulation
Zero-day exploits	Attacks exploiting unknown vulnerabilities.	Unauthorized access, data breach
Eavesdropping	Listening to communication to gather credentials.	Unauthorized access, data breach
XSS	Injecting malicious scripts into web interfaces.	Unauthorized access, data manipulation
DoS	Overwhelming the system to disrupt services.	System downtime, data unavailability
DDoS	Coordinating multiple devices to perform DoS attacks.	System downtime, data unavailability
Side channel attacks	Exploiting system information leakage during authentication.	Unauthorized access, data breach
Tampering	Modifying IoMT device or authentication data.	Data manipulation, system compromise

DDoS, Distributed DoS; *DoS*, denial-of-service; *MITM*, man-in-the-middle; *XSS*, cross-site scripting.

1.4 Technological advancements for authentication in Internet of Medical Things

IoMT has been developing quickly as a result of technology developments targeted at enhancing authentication and security. As demonstrated

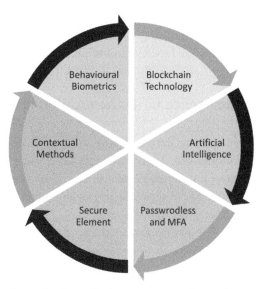

Figure 1.5 Six major technological advancements in the authentication for Internet of Medical Things.

in Fig. 1.5, the six major technological advancements for achieving effective authentication includes blockchain technology, artificial intelligence (AI), behavioral biometrics, passwordless and MFA, secure element (SE) authentication, and contextual authentication.

1.4.1 Blockchain technology for authentication in Internet of Medical Things

Blockchain technology has the ability to completely transform authentication and security in the health-care sector when it is integrated into the IoMT. Immutability, decentralization, and transparency—three fundamental characteristics of blockchain—address the difficult problems of protecting private medical information and guaranteeing the reliability of connected devices. Health-care providers may set up a reliable and impenetrable identity verification system for patients, doctors, and IoT devices by using blockchain-based authentication processes. The blockchain securely stores each transaction or access request, producing an immutable trail that can be checked by numerous network nodes. The potential dangers and vulnerabilities associated with centralized authentication systems are reduced by this strong authentication technique, which guards against unauthorized access, data alteration, and identity theft.

Furthermore, the transparency of blockchain fosters confidence among patients, health-care providers, and other stakeholders, building a more dependable and accountable IoMT framework with enormous potential to alter patient outcomes and health-care delivery [29].

1.4.2 Artificial intelligence for authentication in Internet of Medical Things

To protect sensitive patient data and stop unwanted access, traditional authentication techniques like passwords and PINs are no longer sufficient. To offer creative and dependable authentication solutions, AI enters the picture. The advantages of using AI in implementation authentication are listed next.

- *Behavioral analysis*: AI systems can examine patterns of user behavior, including typing speed, navigation preferences, and device usage. As a result, the system is able to recognize any irregularities or suspicious activity and send out authentication prompts as needed.
- *Continuous authentication*: Conventional authentication procedures frequently entail a one-time login procedure, leaving the system open to attack by an unauthorized user. Continuous authentication driven by AI keeps track of user activity throughout the session and offers a dynamic security layer that changes in response to new threats.
- *Contextual authentication*: When determining the validity of a login attempt, AI can use contextual data such as the user's location, the time of access, and the device's specifications. As a result, the risk of false positives is decreased, and the user experience is enhanced overall.
- *Threat detection and mitigation*: In real time, AI systems can examine network traffic and device interactions to spot potential risks and malicious behavior. IoMT systems can prevent attacks and safeguard patient data by making use of this intelligence.

Although AI has several advantages for IoMT authentication, it also presents problems of data privacy, algorithm robustness, and ethics that need to be solved. Strict privacy controls are needed for the collection and processing of biometric and behavioral data to protect patient data's security and confidentiality. To keep AI models effective against new cyber threats, they must be continually updated and evaluated [30]. When using AI to healthcare, it is important to prioritize patient consent, openness, and ethical issues. The use of AI for authentication in IoMT systems offers a potent approach to improve security, safeguard patient data, and stop illegal access to private medical data. While AI offers cutting-edge capabilities, its implementation must be done responsibly, with an emphasis on privacy, data protection, and ethical norms, to ensure the industry's acceptance of these technologies and its confidence in them [31]. To ensure the integrity and security of connected health-care devices and applications as the IoMT ecosystem develops, AI-powered authentication must be integrated.

1.4.3 Passwordless and multifactor authentication in Internet of Medical Things

Password-based systems are prone to a number of flaws, including social engineering, brute-force attacks, and password reuse. By doing away with passwords, passwordless authentication increases the overall security of IoMT systems [32]. Instead, it makes use of safer techniques, including hardware-based authentication (security keys), token-based verification, and biometric authentication (fingerprint, face, or iris recognition). Only authorized users will be able to access critical medical data thanks to these technologies' inherent strength and difficulty for attackers to penetrate. By requesting several forms of identity from users before granting access to IoMT systems, MFA offers an additional layer of protection. MFA often involves a user's knowing something (like a password), having something (like a smartphone or security token), and being something (like their biometric information) [33]. Critical medical information is protected from cyberattacks thanks to this multilayered strategy, which dramatically lowers the danger of illegal access and identity theft. The advantages of passwordless and MFA are explained next.

- *Stronger security*: By doing away with weak passwords and demanding multiple authentication methods, the risk of successful cyberattacks is greatly decreased.
- *User-friendliness*: Passwordless authentication techniques are more effective and convenient for users, which lowers the likelihood of password-related problems like forgotten passwords or the requirement for frequent password resets.
- *Compliance and regulation*: Strict data protection laws, such as HIPAA in the United States, must be followed by the health-care sector. Compliance with these rules is guaranteed by the use of strong authentication techniques like passwordless and MFA.
- *Reduced fraud*: In the IoMT ecosystem, passwordless and MFA protocols create an extra degree of defense against fraud and medical identity theft.

- *Enhanced trust*: Patients and health-care professionals can both have more faith in the security of IoMT systems, which will encourage them to have more faith in connected medical services and equipment.

The necessity for sophisticated security solutions is increasingly important as the IoMT landscape develops [34]. Securing the integrity of medical data and protecting patient privacy have advanced significantly with passwordless authentication and MFA. By embracing these technologies, health-care organizations are not only shielded from security flaws but are also helping to create a more secure and dependable environment for the development of connected medical services and devices in the future. The IoMT community can accept innovation with confidence if security is prioritized, which will be advantageous for patients and caregivers around the world [35].

1.4.4 Secure element (hardware) authentication in Internet of Medical Things

To improve the security and privacy of medical equipment and systems, the IoMT uses SE authentication. An SE is a specialized hardware piece created to store and handle sensitive data and cryptographic keys in a secure manner [36]. Making it impossible for attackers to tamper with or extract critical information, it offers a safe environment that is segregated from the rest of the device's operating system and hardware. SEs are frequently found in smart cards, cell phones, payment systems, and other gadgets that call for strong security measures. Although SEs greatly improve security, it is vital to remember that other security measures should also be used in IoMT. To build a strong and secure IoMT ecosystem, a holistic approach to security is required, including secure communication protocols, frequent software updates, vulnerability analyses, and tight access controls. Additionally, security should be incorporated into the system from the beginning to ensure that it is complete and effective.

1.4.5 Contextual authentication in Internet of Medical Things

The practice of confirming the identity of people or devices inside the health-care ecosystem using contextual information and variables is known as contextual authentication in the IoMT field. Contextual authentication aims to boost the security of IoMT systems while preserving a smooth and user-friendly experience for medical staff and patients [37]. In the complex and dynamic IoMT environment, conventional authentication techniques like passwords or PINs might not be sufficient.

To decide which access to grant, contextual authentication uses additional contextual data. Contextual authentication is a powerful security tool that may be implemented in IoMT systems to increase security while lowering the possibility of hacking or data leaks. Because it enables adaptive security measures, a more effective defense against potential threats is offered. These measures can change depending on the context of each access attempt. Balancing security and user experience, though, is absolutely essential. Excessive security measures could potentially affect patient care by making health-care personnel less effective and inconvenient. Contextual authentication solutions must therefore be carefully created and fine-tuned by IoMT administrators and developers to guarantee both security and usability in medical settings.

1.4.6 Behavioral biometrics for authentication in Internet of Medical Things

Behavioral biometrics refers to the use of distinctive behavioral patterns and traits of individuals to confirm their identification and give access to private medical equipment or data within the IoMT ecosystem. The IoMT is a network of linked medical devices and systems that gather, monitor, and transmit health-care data; therefore it is critical to have trustworthy and secure authentication processes in place to safeguard patient data and guard against illegal access [38]. Passwords and PINs, which are common forms of authentication, can be inconvenient and open to security flaws. Behavioral biometrics, on the other hand, makes advantage of a person's distinctive behavioral traits and behaviors to develop a more reliable and approachable authentication procedure. For IoMT authentication, some frequent behavioral biometric factors are listed next.

- *Keystroke dynamics*: Observing how a user types on a keyboard or touchscreen can reveal idiosyncratic characteristics about that person. The creation of a user profile for authentication takes into account elements like typing speed, keystroke duration, and rhythm.
- *Gait recognition*: Gait analysis entails watching a person's gait, which can be utilized to recognize them individually. IoMT wearables or other devices used in ambulatory settings may find this to be very helpful.
- *Signature verification*: Verifying a person's signature when they use electronic signature pads or touchscreen devices can add an extra degree of authenticity in a digital setting.

- *Voice recognition*: When logging onto IoMT devices with voice interfaces, analyzing speech patterns and voice characteristics might aid to confirm a user's identification.
- *Touch and gesture recognition*: How a person interacts with touchscreens or gestures on devices can potentially be used as a behavioral biometric element.
- *Eye movement*: In some circumstances the user's eye movement, such as gaze direction, can be employed as a biometric factor for identification.

Behavioral biometrics is advantageous for IoMT authentication in a number of ways. Compared to conventional authentication systems, they are less obtrusive and more user-friendly [39,40]. These biometrics may adjust to changes over time, such as aging or injuries, because they are based on behavioral patterns, making them more dependable for long-term use. Nevertheless, it is crucial to resolve any potential privacy issues before integrating behavioral biometrics into IoMT systems. To guarantee that the gathered biometric data are safely preserved and cannot be exploited, appropriate data protection procedures should be in place. In general, behavioral biometrics presents a technique that has promise for enhancing the security and UI of IoMT devices while protecting patient data. The application of behavioral biometrics in health-care settings is predicted to grow in popularity as technology develops.

1.5 Conclusion

IOMT is a network of hardware infrastructures, software applications, and clinical equipment connected to the Internet and is a system for integrating health-care information technology systems. Protecting sensitive and personal data is one of the main security issues for wearable and implantable devices in the IoMT setting. The most crucial security component in the IoMT system is authentication since, when utilized correctly, it may effectively thwart various threats. The authentication process of the IoMT system is covered in detail in this chapter. The explanation has been expanded to include the security criteria of authentication schemes, a taxonomy of authentication protocols, and a list of the current authentication schemes. The chapter also looks at the technology developments used to create the authentication protocols for IoMT, including blockchain, AI, passwordless authentication, MFA, SE authentication, contextual authentication, and behavior biometric technologies.

References

[1] A.K. Singh, B.D.K. Patro, Performance comparison of signcryption schemes—a step towards designing lightweight cryptographic mechanism, International Journal of Engineering and Technology (IJET) 9 (2) (2017) 1163–1170.

[2] R. Hireche, H. Mansouri, A.-S. Khan Pathan, Security and privacy management in Internet of Medical Things (IoMT): A synthesis, Journal of Cybersecurity and Privacy 2 (3) (2022) 640–661.

[3] J.-P.A. Yaacoub, M. Noura, H.N. Noura, O. Salman, E. Yaacoub, R. Couturier, et al., Securing Internet of Medical Things systems: limitations, issues and recommendations, Future Generation Computer Systems 105 (2020) 581–606.

[4] S.-A. Ayoub, M.A. Al-Garadi, N. Boustia, Survey of machine learning based intrusion detection methods for Internet of Medical Things, Applied Soft Computing 140 (2023) 110227.

[5] A. Khan, A. Ahmad, M. Ahmed, J. Sessa, M. Anisetti, Authorization schemes for internet of things: requirements, weaknesses, future challenges and trends, Complex & Intelligent Systems 8 (5) (2022) 3919–3941.

[6] A.K. Singh, A. Nayyar, A. Garg, A secure elliptic curve based anonymous authentication and key establishment mechanism for IoT and cloud, Multimedia Tools and Applications 82 (2023) 22525–22576.

[7] M. Papaioannou, M. Karageorgou, G. Mantas, V. Sucasas, I. Essop, J. Rodriguez, et al., A survey on security threats and countermeasures in Internet of Medical Things (IoMT), Transactions on Emerging Telecommunications Technologies 33 (6) (2022) e4049.

[8] N. Alsaeed, F. Nadeem, Authentication in the Internet of Medical Things: taxonomy, review, and open issues, Applied Sciences 12 (15) (2022) 7487.

[9] P. Porambage, C. Schmitt, P. Kumar, A. Gurtov, M. Ylianttila, Two-phase authentication protocol for wireless sensor networks in distributed IoT applications, in: 2014 IEEE Wireless Communications and Networking Conference (WCNC), IEEE, 2014, pp. 2728–2733.

[10] S. Agrawal, P. Ahlawat, A survey on the authentication techniques in internet of things, in: 2020 IEEE International Students' Conference on Electrical, Electronics and Computer Science (SCEECS), IEEE, 2020, pp. 1–5.

[11] A. Singh, Kumar, B.D.K. Patro, A novel security protocol for wireless sensor networks based on elliptic curve signcryption, International Journal of Computer Networks & Communications (IJCNC) 11 (2019) 93–112.

[12] H. Amintoosi, M. Nikooghadam, M. Shojafar, S. Kumari, M. Alazab, Slight: a lightweight authentication scheme for smart healthcare services, Computers and Electrical Engineering 99 (2022) 107803.

[13] A.K. Singh, B.D.K. Patro, Signcryption-based security framework for low computing power devices, Recent Advances in Computer Science and Communications (Formerly: Recent Patents on Computer Science) 13 (5) (2020) 845–857.

[14] A.K. Singh, A. Solanki, A. Nayyar, B. Qureshi, Elliptic curve signcryption-based mutual authentication protocol for smart cards, Applied Sciences 10 (22) (2020) 8291.

[15] K.-H. Yeh, A secure IoT-based healthcare system with body sensor networks, IEEE Access 4 (2016) 10288–10299.

[16] P. Gope, T. Hwang, A realistic lightweight anonymous authentication protocol for securing real-time application data access in wireless sensor networks, IEEE Transactions on Industrial Electronics 63 (11) (2016) 7124–7132.

[17] Z. Zhao, An efficient anonymous authentication scheme for wireless body area networks using elliptic curve cryptosystem, Journal of Medical Systems 38 (2014) 1–7.

[18] P. Kumar, S.-G. Lee, H.-J. Lee, E-SAP: efficient-strong authentication protocol for healthcare applications using wireless medical sensor networks, Sensors 12 (2) (2012) 1625–1647.

[19] D. He, N. Kumar, J. Chen, C.-C. Lee, N. Chilamkurti, S.-S. Yeo, Robust anonymous authentication protocol for health-care applications using wireless medical sensor networks, Multimedia Systems 21 (2015) 49–60.

[20] X. Li, J. Niu, S. Kumari, J. Liao, W. Liang, M.K. Khan, A new authentication protocol for healthcare applications using wireless medical sensor networks with user anonymity, Security and Communication Networks 9 (15) (2016) 2643–2655.

[21] Z.U. Rehman, S. Altaf, S. Iqbal, An efficient lightweight key agreement and authentication scheme for WBAN, IEEE Access 8 (2020) 175385–175397.

[22] L. Mucchi, L.S. Ronga, L. Cipriani, A new modulation for intrinsically secure radio channel in wireless systems, Wireless Personal Communications 51 (2009) 67–80.

[23] S. Soderi, L. Mucchi, M. Hämäläinen, A. Piva, J. Iinatti, Physical layer security based on spread-spectrum watermarking and jamming receiver, Transactions on Emerging Telecommunications Technologies 28 (7) (2017) e3142.

[24] M. Ibrahim, S. Hamada, A.K. Kumari, M. Das, Wazid, V. Odelu, Secure anonymous mutual authentication for star two-tier wireless body area networks, Computer Methods and Programs in Biomedicine 135 (2016) 37–50.

[25] Z. Xu, C. Xu, W. Liang, J. Xu, H. Chen, A lightweight mutual authentication and key agreement scheme for medical Internet of Things, IEEE Access 7 (2019) 53922–53931.

[26] J. Zhao, H. Hu, F. Huang, Y. Guo, L. Liao, Authentication technology in internet of things and privacy security issues in typical application scenarios, Electronics 12 (8) (2023) 1812.

[27] M.K. Hasan, T.M. Ghazal, R.A. Saeed, B. Pandey, H. Gohel, A.'A. Eshmawi, et al., A review on security threats, vulnerabilities, and counter measures of 5G enabled Internet-of-Medical-Things, IET Communications 16 (5) (2022) 421–432.

[28] A.K. Singh, B.D.K. Patro, Security attacks on RFID and their countermeasures, in: Computer Communication, Networking and IoT: Proceedings of ICICC 2020, Springer Singapore, 2021, pp. 509–518.

[29] F. Fotopoulos, V. Malamas, T.K. Dasaklis, P. Kotzanikolaou, C. Douligeris, A blockchain-enabled architecture for IoMT device authentication, in: 2020 IEEE Eurasia Conference on IoT, Communication and Engineering (ECICE), IEEE, 2020, pp. 89–92.

[30] F. Muheidat, L.A. Tawalbeh, AIoMT artificial intelligence (AI) and Internet of Medical Things (IoMT): applications, challenges, and future trends, in: Y. Maleh, A. A. Abd El-Latif, K. Curran, P. Siarry, N. Dey, A. Ashour, S.J. Fong (Eds.), Computational Intelligence for Medical Internet of Things (MIoT) Applications, Academic Press, 2023, pp. 33–54.

[31] P. Manickam, S.A. Mariappan, S.M. Murugesan, S. Hansda, A. Kaushik, R. Shinde, et al., Artificial intelligence (AI) and Internet of Medical Things (IoMT) assisted biomedical systems for intelligent healthcare, Biosensors 12 (8) (2022) 562.

[32] M.J. Haber, B. Hibbert, Passwordless authentication, Privileged Attack Vectors: Building Effective Cyber-Defense Strategies to Protect Organizations, O'Reilly, 2020, pp. 87–98.

[33] T. Suleski, M. Ahmed, W. Yang, E. Wang, A review of multi-factor authentication in the Internet of Healthcare Things, Digital Health 9 (2023). 20552076231177144.

[34] A. Garg, A.K. Singh, Internet of Things (IoT): security, cybercrimes, and digital forensics, in: K. Kaushik, S. Dahiya, A. Bhardwaj, Y. Maleh (Eds.), Internet of Things and Cyber Physical Systems, CRC Press, 2022, pp. 23–50.

[35] A. Garg, A.K. Singh, Applications of Internet of Things (IoT) in green computing, in: F. Al-Turjman, A. Nayyar, A. Devi, P. K. Shukla (Eds.), Intelligence of Things: AI-IoT based critical-applications and innovations, Springer, 2021.

[36] C. Patel, A.K. Bashir, A.A. AlZubi, R. Jhaveri, EBAKE-SE: a novel ECC-based authenticated key exchange between industrial IoT devices using secure element, Digital Communications and Networks 9 (2) (2023) 358−366.

[37] Y. Ashibani, D. Kauling, Q.H. Mahmoud, Design and implementation of a contextual-based continuous authentication framework for smart homes, Applied System Innovation 2 (1) (2019) 4.

[38] A. Garg, A.K. Singh, Applications of Internet of Things (IoT) in green computing, in: F. Al-Turjman, A. Nayyar, A. Devi, P. K. Shukla. Intelligence of Things: AI-IoT Based Critical-Applications and Innovations, Springer International Publishing, 2021, pp. 1−34.

[39] M. Nerini, E. Favarelli, M. Chiani, Augmented PIN authentication through behavioral biometrics, Sensors 22 (13) (2022) 4857.

[40] P.M.A.B. Estrela, R.de O. Albuquerque, D.M. Amaral, W.F. Giozza, R.T.de S. Júnior, A framework for continuous authentication based on touch dynamics biometrics for mobile banking applications, Sensors 21 (12) (2021) 4212.

CHAPTER 2

Optimal machine learning—based data classification on Internet of medical things environment

P. Maheswaravenkatesh[1], A.N. Arun[2] and T. Jayasankar[1]
[1]Department of Electronics and Communication Engineering, Anna University, Trichy, Tamil Nadu, India
[2]Department of Computer Science & Engineering, Sri Venkateswara Institute of Science and Technology, Tiruvallur, Tamil Nadu, India

2.1 Introduction

Smart healthcare offers healthcare platforms that exploit gadgets like wearables and mobile Internet to conveniently enter health documents and connect resources, individuals, and organizations. A smart, networked medical device, called the Internet of Medical Things (IoMT), has connected people worldwide. This enables the monitoring of a large variety of unknown healthcare data [1]. The need for healthcare data, mainly visual depictions of health, like images and signals, has been raised recently. IoT applications, which include remote patient monitoring, wearables, prescription tracking system, and network for the medical supply chain, were broadly leveraged in the healthcare sectors [2]. IoM helps doctors in providing very accurate diagnoses by upholding a permanent record of the current health conditions of patients [3]. Patents may interact with their nurses and clinicians through smartphone applications. It makes it possible for doctors to treat many patients quickly.

The IoT technology tracks the location of hospitalized patients, presents virtual care for patients affected by long-term diseases, monitors the treatment of patients, and can render data to caregivers and portable health gadgets for patients [4]. The IoMT technology saves the efforts and time of doctors and patients. It connects patients with their medical practitioners and facilitates the secure transfer of healthcare data through a network, thereby minimizing problems in healthcare systems [5]. A rapid rise in the growth and utility of IoMT paves the way to deploy these structures that can accurately, rapidly, and securely scrutinize patients' health and identify and cure various distantly. IoT-related structures were abundant, mainly for

Securing Next-Generation Connected Healthcare Systems
DOI: https://doi.org/10.1016/B978-0-443-13951-2.00001-5

diseases that were very significant in patients' life, like cardiovascular disease [6]. In medical research, heart disease (HD) prediction is challenging and depends on observing numerous symptoms involving shortness of breath, chest pain, blood pressure, cold sweats, chest congestion, etc. IoT sensor values were considered input to forecast HD and assist diagnosis [7]. Historically, diseases have been identified by clinical and physical examination. Nowadays, a smartwatch helps detect health irregularities, for instance, an irregular heartbeat in elderly persons [8].

Commonly, medical data are in the form of electronic medical reports gathered from patients [9]. Mostly the data in healthcare were used to build a decision support system, which uses patient data with artificial intelligence and field knowledge. These techniques, often constructed into clinical decision support systems, are diagnostic methods related to machine learning (ML) that forecast the incidence of a disease in a patient related to a set of risk features [10]. Though ML methods were broadly studied and seem to be successful, heart-disease estimation was a complex issue and there are still several enhancements that should be done and techniques that must be explored.

This study presents a Deep Equilibrium Optimizer based Feature Selection with Optimal Machine Learning for heart disease diagnosis (DEOFSOML-HDD) in the IoMT platform. The DEOFSOML-HDD technique enables the IoMT devices to gather patient data for disease detection. Data preprocessing is performed to convert the raw medical data into a useful format. In addition, the DEOFS technique is exploited to optimally choose the subset of features. Moreover, the presented DEOFSOML-HDD technique utilizes a modified beetle antenna search algorithm (MBASA) with a variational autoencoder (VAE) for disease detection. An extensive set of simulations are carried out to examine the performance of the DEOFSOML-HDD technique.

2.2 Related works

In Ref. [11], the presented recursion enhanced RF with an enhanced linear model (RFRF-ILM) for diagnosing HD. This study identifies the main features of estimating HDs using ML approaches. The predictive technique adds many feature combinations and several established techniques of classification. It would produce a superior level of performance with accuracy by using HD predictive techniques. In his work, the elements causing HD were identified. In Ref. [12], the smart healthcare technique was modeled for predicting HD through a biogeography optimization algorithm and Mexican hat wavelet to enhance dragonfly algorithm optimization with a mixed KELM approach. In his study, data were collected from the two gadgets, namely, sensor nodes, along with the electronic medical reports. The android-related model was employed for collecting the data of patients and the reliable cloud-related method for data storage.

Lu et al. [13] presented a long short term memory (LSTM) network to enhance it. Ac classifier method arithmetically related to the CNN-LSTM network method was modeled. At first, the deep CNN was devised for encoding electrocardiography (ECG) signals and deriving morphological features of ECG signals. Then, intrinsic features were deeply extracted by using the temporal relation of LSTM learning morphological feature representation. Nin et al. [14] devised an automatic CHF detection method related to the hybrid DL method comprising an recurrent neural network (RNN) and convolutional neural network (CNN). The author even categorized CHF and normal sinus heart rate signals depending on time-frequency spectra and ECG at the time of the RR interval. Mastoi et al. [15] introduced a new healthcare-related fog cloud system (HCBFS) for determining, collecting, and analyzing critical tasks to diminish the total cost.

Abdel-Basset et al. [16] proposed a new structure relevant to computer IoT and supported diagnosis for monitoring and detecting heart failure—infected patients. The data can be acquired from several sources. This presented a healthcare mechanism focused on better diagnosis accuracy with unclear data. The author devised a neutrosophic multicriteria decision-making approach to help doctors and patients know whether a patient suffers from heart failure. Reddy et al. [17] purpose of the adaptive genetic algorithm (GA) with fuzzy logic (FL), named (AGAFL) method was to forecast HD, which assists doctors in identifying HD in the initial phase. The generated rules from fuzzy techniques were by implementing adaptive GA. Initially, significant features that affect cardiovascular disease were chosen by rough set theory. The next step estimates the HD utilizing the hybrid AGAFL technique.

2.3 The proposed model

This study presented an IoMT-related framework for enhancing the prediction of HD. Fig. 2.1 illustrates the block diagram of the DEOFSOML-HDD system. The IoMT gadget gathers patient data relating to the heart after and before the arrival of HD [18]. The patients' health parameters

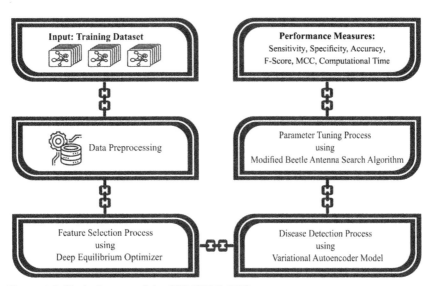

Figure 2.1 Block diagram of the DEOFSOML-HDD system.

were tracked distantly, constantly, in real-time, and after being saved and transferred to the data center, like the cloud, which pointedly raises the accessibility, cost-effectiveness, and efficacy of the healthcare system. Sensor data gathered to diagnose the patient's heart condition were sent to the hospital administration. Testing and training stages were carried out to determine the patient's heart condition. The UCI data repositories are leveraged for training the data values, preprocessing, selecting features utilizing the EOFS method, and classification utilizing MBASA with VAE. The final classifier outcomes specify whether the patient's heart condition was abnormal or normal. Depending on these outcomes, the required initiatives are taken by the clinician.

2.3.1 Preprocessing

Primarily, the medicinal data have to be preprocessed in two manners: data normalization and missing value replacement. The data value was collected in the UCI database, whereas it can be preprocessing that the missing value replacement or needs noise removal [19]. The noiseless data support an effective identifying pattern compared with HDs. The median studentized remaining approach was utilized to eliminate noisy or unwanted data as it inspects the relations Betwixt data in the databases. This process of noise reduction enhances the recognition method of HDs.

2.3.2 Feature selection using DEO algorithm

Once the input data have been preprocessed, the DEOFS approach is exploited to choose the subset of features optimally. DEOFS is developed for selecting the optimum combination of features. Equilibrium optimization algorithm (EOA) was established by using the equilibrium of dynamic mass for control volumes by looking for a balanced model state to overcome optimized issues [20]. Even with the provided advantages, EO has the shortcomings of lacking interest in fitness assignment, and it is not capable to fulfill the contrary objective carried by multiobjective function because of its higher propensity of accomplishing equilibrium. To solve the disadvantages of EOA, the study is developed to manage optimum FS, usually expressed as multiobjective optimization problems. The presented method uses a hyperlearning technique which could employ the concept of personal worst and best states in solution upgrading procedures to resolve the multiobjective FS problems. Without losing generalization, the presented method comprises n subswarms of particle concentration (position) vector represented by X to final optimal features subset by representing the fitness value of the candidate solution. Compared with single-swarm EOA, the presented method leverages the benefits of an external shared pool to enable sharing of equilibrium state experiences among diverse sub-swarms, enabling the particle to approach the exact Pareto front very efficiently. The search procedure of the presented method is defined in the following. At the beginning of the optimized technique, the first location of every i-th particle, namely, X_i for $i = 1, 2, \ldots, n_p$, is produced in the following:

$$X_{initial} = rand(n_p, d) \times (ub - lb) + lb \qquad (2.1)$$

where n_p indicates population size, d denotes dimensional size; lb and ub signify the lower and upper bounds of decision variables, correspondingly. Aft rewards finishing the initialization method, the average position $(X_{eq,av})$ of the population and four best equilibrium particles $(X_{eq,1}, X_{eq,2}, X_{eq,3}, X_{eq,4})$ are recognized to create an equilibrium pool X_{eq}. This pool offers many promising search patterns in the following:

$$X_{eq,pool} = (X_{eq,1}, X_{eq,2}, X_{eq,3}, X_{eq,4}, X_{eq,av}) \qquad (2.2)$$

For all the iterations, the original location X_{old} of all the particles in each subswarm is upgraded by communicating with the solution member X_{eq} arbitrarily chosen from the equilibrium pool $X_{eq,pool}$. The following expression accomplishes the solution updating mechanism of all the particles:

$$X_{new} = X_{eq} + \frac{G}{\lambda}(1 - F) + \left(X_{old} - X_{eq}\right) \times F \qquad (2.3)$$

$$F = a_1 \, sign(r - 0.5)\left(e^{-\lambda t} - 1\right) \qquad (2.4)$$

where X_{old} and X_{new} indicate the existing and novel location vectors of the particle, correspondingly, r denotes the random value within 0 and 1; $a1$ shows constant ($a1 = 2$); λ was a random vector with values from zero to one:

$$t = \left(1 - \frac{T}{T_{max}}\right)^{a_2 \left(\frac{T}{T_{max}}\right)} \qquad (2.5)$$

where a_2 denotes constant ($a_2 = 1$), T_{max} and T symbolize the maximal iteration amount and existing iteration count, correspondingly:

$$G = \begin{cases} 0.5r_1 & if \quad r_2 \geq GP \\ 0 & if \quad r2 < GP \end{cases} \qquad (2.6)$$

where r_1 and r_2 denote the random number ranges from zero to one, GP represents a generation probability and is fixed to 0.5. The presented method can be modified from EOA by integrating external archive dominance criteria for finding a suitable solution to tackle the multi-objective optimization problem that is usually defined by:

$$Min F(X) = \left\{f_1(x), f_2(x), \dots, f_n(x)\right\}$$

$$\text{Subjected to:} \begin{cases} g_i(X) \leq 0 i = 1, 2, \dots - q \\ h_i(X) = 0 i = 1, 2, \dots l \end{cases} \qquad (2.7)$$

where $F(x)$ indicates the vector of multiobjective functions, $g_i(X)$ and $h_i(X)$ represent q and l inequality and equality constraints, correspondingly.

2.3.3 Optimal variational autoencoder–based classification

The VAE model is used in this study for the disease detection process. VAEs get the infrastructure of deep AEs, then impose further constraints on the bottleneck which transforms typical deterministic AEs into great probabilistic methods [21]: but deep AEs learn a random purpose to encode and decoded input data, VAE learns the parameter of probability distributions which data models. The VAE contains encoded $q_\varphi(z|x)$, being a nearby posterior, and decoded $p_\theta(x|z)$, the present possibility of data x to provide latent variable z. Based on this approach, we develop an encoded variational inference network to map input data for approximating the posterior distribution in the latent space, and decodes functions as a generative network, mapping random latent coordinate back to distribution on novel data space. It can be considered which input data have been sampled in a unit Gaussian distribution of latent variables. The learn procedure method was trained as concurrently optimized two loss function the reconstruction loss $\mathcal{L}$ and Kullback−Leibler divergence D_{KL} Betwixt learned latent distribution and prior unit Gaussian. Before it could understand VAE as deep AE with a further regularization offered by the D_{KL} term, Fig. 2.2 showcases the infrastructure of VAE.

The resultant main function of VAE has been projected, and it can be variational lower bound of the marginal possibility of data, while the marginal possibility was intractable. The marginal probability is the sum above the marginal probability of different data points $\log p_\theta(x) = \sum_{i=1}^{n} \log p_\theta(x^{(i)})$, and it could be modified to separate data points $x^{(i)}$ as:

$$\log p_\theta(x^{(i)}) = D_{KL}(q_\varphi(z|x^{(i)})||p_\theta(z|x^{(i)})) + \mathcal{L}(\theta, \varphi; x^{(i)}) \qquad (2.8)$$

Assuming the D_{KL} term was constantly superior to 0, and executing Bayes rule, variational lower bound $\log p_\theta(x^{(i)})$ outcomes that:

$$-D_{KL}(q_\varphi(z|x^{(i)})||p_\theta(z)) + E_{q_\varphi}(z|, x^{(i)}) \left[\log p_\theta(x^{(i)}|, z) \right] \qquad (2.9)$$

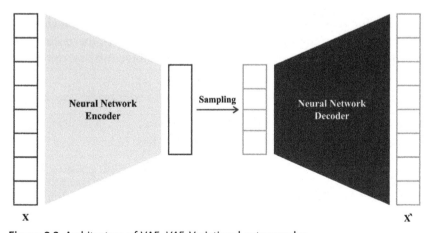

Figure 2.2 Architecture of VAE. *VAE*, Variational autoencoder.

In the learning procedure, VAE has trained for maximizing, utilizing the gradient descent by back-propagation (BP). However, this causes some problems, as the reconstruction error term in Eq. (2.8) needs a Monte Carlo estimation of expectations that could not simply be differentiable. To exceed that, the *VAE* comprises a reparameterization trick that utilizes an arbitrary variable in a typical normal distribution rather than an arbitrary variable in the novel distribution $(z = \mu + \sigma e)$, with $e \sim N(0, 1)$. This trick permits propagating the gradient back with the network and executing the BP system for training the VAE. This approach can be attained through AE, which works as a generative method.

Finally, the MBASA is applied to adjust the parameters related to the VAE model. BAS mimics the natural behavior of the beetle's antennae [22]. But the beetle antennae search technique leverages single beetle, which was hard to find the optimum parameter while the range of parameter to be improved are wider. Before MBASA applied multibeetles, raising the probability of attaining the optimum parameter, the procedure of parameter selection of VAE based on MBASA is discussed in the following:

Step 1: Describe beetles' position as a vector x^t at the *t-th* time instant $(t = 1, 2 \cdots)$. Initialize parameter of MBASA, involving step size δ^0, the beetles' location x^0, and antennae length d^0.

Step 2: Assess the fitness of every beetle fivefold cross-validation technique was exploited to estimate the fitness of the beetle. In the fivefold cross-validation technique, the training sample is divided equally into five subsets of samples, among them, four subsets of samples were leveraged for training the VAE, and the residual set was leveraged for testing the VAE. Sequentially, every subset is utilized as the testing set. The overall diagnoses performance A_i of five subsets of the samples is attained by:

$$A_j = \frac{N_{correct,i}}{n\ total} \tag{2.10}$$

The fitness of *i-th* beetles can be determined by:

$$f(x_i) = 1 - A_j \tag{2.11}$$

Step 3: Attain the search behavior of the right- and left-hand sides.

For modeling the searching behaviors, a random direction of beetle search is defined by the following expression:

$$\vec{b} = \frac{rv(k, 1)}{eps + ||rv(k, 1)||} \tag{2.12}$$

where $rv(m, 1)$ indicates an m-dimension vector with random value amongst $[-1,1]$, m refers to the position dimension, now, m is fixed to 2, and $eps = 2^{-52}$.

The activity of the beetles' antennae was imitated by the search behavior of the right- and left-hand sides:

$$\begin{cases} x_{i,r} = x_i^t + d^t\vec{b} \\ x_{i,l} = x_i^t - d^t\vec{b} \end{cases} \tag{2.13}$$

where $x_{i,r}$ indicates the location in the search region of the i-th beetle's right-hand side, $x_{i,l}$ denotes the location in the search region of the i-th beetle's left-hand side. It shows the sensing extent of antennae respective to the exploit capability at this time.

Step 4: Upgrade the position of the beetles.

The iterative method can be produced as Eq. (2.13) to related with the odor recognition by taking the searching activity into account,

$$x_i^t = x_i^{t-1} + \delta^t\vec{b}sign\left(f\left(x_{i,r}\right) - f\left(x_{i,l}\right)\right) \tag{2.14}$$

Δt indicates the step size of searching, and a sign shows a sign function.

Step 5: Compare the fitness of x_i^t with the fitness of the existing best location of i-th beetles; when $f(x_i^t) < f_{best}$, then $f_{best} = f(x_i^t)$, $x_{i,best} = x_i^t$, whereas $x_{i,best}$ indicates the present optimum location of the i-th beetles, and $f_{i,best}$ shows the fitness of present optimum location of i-th beetles.

Step6: Upgrade the antennae length d and step size δ.

$$d^t = 0.95d^{t-1} + r0 \tag{2.15}$$

$$\delta^t = 0.95\delta^{t-1} \tag{2.16}$$

Whereas $r0$ shows constantly.

Step 7: Repeat steps 2−6 until the ending condition is reached.

Step 8: Attain the optimum location of the best fitness amongst each beetle which is the optimum parameter of VAE.

2.4 Experimental validation

In this section, the HD classification results of the DEOFSOML-HDD model are investigated using a dataset consisting of 303 samples as depicted in Table 2.1, each of which has two classes and 13 features (age, sex, cp, restbps, chol, fbs, restecg, thalach, examg, and oldpeak).

Fig. 2.3 illustrates the best cost (BC) investigation of the DEOFSOML-HDD model with other FS methods. The results implied that the DEOFSOML-HDD method had shown an optimal BC of 0.4145 with six features.

The confusion matrices attained by the DEOFSOML-HDD model under several epoch counts are reported in Fig. 2.4. On 00 epochs, the DEOFSOML-HDD model has categorized 138 samples into the present and 119 samples into absent classes. In addition, on 300 epochs, the DEOFSOML-HDD approach has categorized 155 samples into present and 128 samples into absent classes. Additionally, on 500 epochs, the DEOFSOML-HDD technique has categorized 153 samples into present and 130 samples into absent classes.

Table 2.2 outlines the HD detection performance of the DEOFSOML-HDD model under diverse epochs. The experimental values highlighted that the DEOFSOML-HDD model had proficiently identified the presence and absence of HD. For instance, with 100 epochs, the DEOFSOML-HDD model has attained an average $accu_{bal}$ of 84.88%, $sens_y$ of 84.88%, $spec_y$ of 84.88%, F_{score} of 84.76%, and MCC of 69.58%, whereas, with 300 epochs, the DEOFSOML-HDD technique has achieved an average $accu_{bal}$ of 93.30%, $sens_y$ of 93.30%, $spec_y$ of 93.30%, F_{score} of 93.35%, MCC of 86.70%. Eventually, with 400 epochs, the DEOFSOML-HDD methodology has reached an average $accu_{bal}$ of 90.86%, $sens_y$ of 90.86%, $spec_y$ of 90.86%, F_{score} of 90.72%, MCC of 81.52%. Finally, with 500 epochs, the DEOFSOML-HDD approach has achieved an average $accu_{bal}$ of 93.41%, $sens_y$ of 93.41%, $spec_y$ of 93.41%, F_{score} of 93.36%, and MCC of 86.73%.

Table 2.1 Dataset details.

Class	No. of samples
Present	164
Absent	139
Total number of samples	303

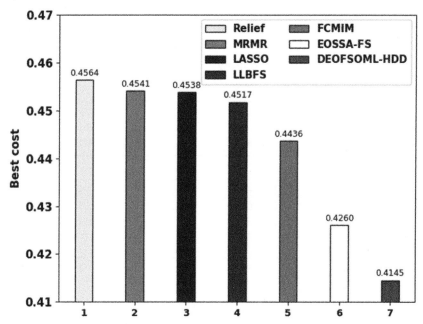

Figure 2.3 BC analysis of DEOFSOML-HDD approach with FS methods. *BC*, Best cost.

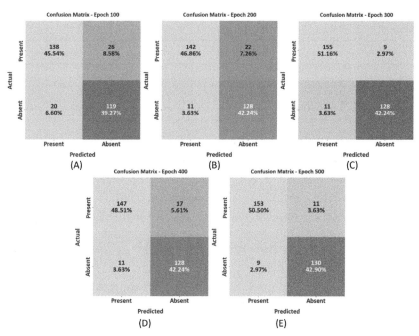

Figure 2.4 Confusion matrices of DEOFSOML-HDD system (A) Epoch100, (B) Epoch200, (C) Epoch300, (D) Epoch400, and (E) Epoch500.

Table 2.2 Heart disease detection outcome of DEOFSOML-HDD system with distinct measures and epochs.

Class	Accuracy$_{bal}$	Sensitivity	Specificity	F-score	MCC
Epoch–100					
Present	84.15	84.15	85.61	85.71	69.58
Absent	85.61	85.61	84.15	83.80	69.58
Average	84.88	84.88	84.88	84.76	69.58
Epoch–200					
Present	86.59	86.59	92.09	89.59	78.41
Absent	92.09	92.09	86.59	88.58	78.41
Average	89.34	89.34	89.34	89.09	78.41
Epoch–300					
Present	94.51	94.51	92.09	93.94	86.70
Absent	92.09	92.09	94.51	92.75	86.70
Average	93.30	93.30	93.30	93.35	86.70
Epoch–400					
Present	89.63	89.63	92.09	91.30	81.52
Absent	92.09	92.09	89.63	90.14	81.52
Average	90.86	90.86	90.86	90.72	81.52
Epoch–500					
Present	93.29	93.29	93.53	93.87	86.73
Absent	93.53	93.53	93.29	92.86	86.73
Average	93.41	93.41	93.41	93.36	86.73

The training accuracy (TACC) and validation accuracy (VACC) of the DEOFSOML-HDD technique are examined on HDD performance in Fig. 2.5. The figure implied that the DEOFSOML-HDD method had shown improved performance with increased values of TACC and VACC. It is noted that the DEOFSOML-HDD algorithm has reached maximum TACC outcomes.

The TLS and VLS of the DEOFSOML-HDD approach are tested on HDD performance in Fig. 2.6. The figure inferred that the DEOFSOML-HDD technique had revealed better performance with minimal values of TLS and VLS. It is noted that the DEOFSOML-HDD technique has resulted in reduced VLS outcomes.

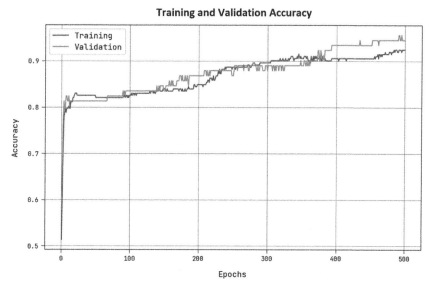

Figure 2.5 TACC and VACC analysis of the DEOFSOML-HDD system.

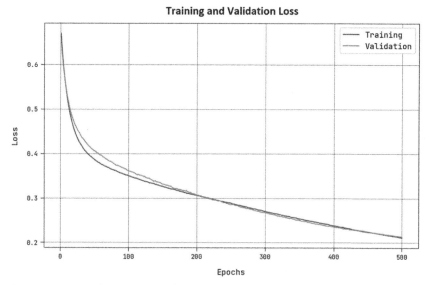

Figure 2.6 TLS and VLS analysis of the DEOFSOML-HDD system.

A clear precision—recall examination of the DEOFSOML-HDD approach under varying epochs is given in Fig. 2.7. The figure indicates the DEOFSOML-HDD method has enhanced values of precision—recall values under distinct epochs.

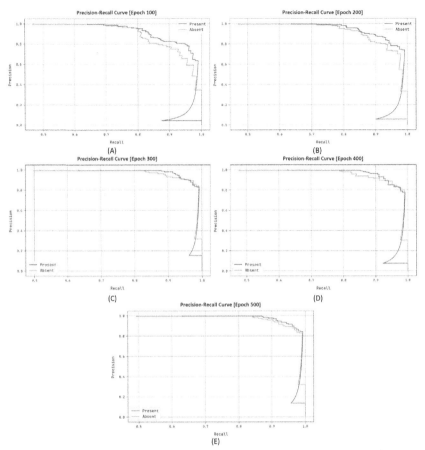

Figure 2.7 Precision–recall of DEOFSOML-HDD system (A) Epoch100, (B) Epoch200, (C) Epoch300, (D) Epoch400, and (E) Epoch500.

A brief receiver operating characteristic (ROC) analysis of the DEOFSOML-HDD methodology under distinct epochs is shown in Fig. 2.8. The outcomes denoted the DEOFSOML-HDD approach has shown its ability to classify different classes under varying epochs.

Table 2.3 reports detailed classification results of the DEOFSOML-HDD model with recent ML models [19]. The simulation values implied that the DEOFSOML-HDD method had gained maximum classifier results with minimal computation time. The comparative study of the DEOFSOML-HDD method with existing ML methods in terms of $accu_y$ is displayed in Fig. 2.9. The results exhibited that the DT and ANN models have obtained reduced $accu_y$ of 79.67% and 75.08%, respectively. The LOR, NB, and KNN models have attained moderately closer $accu_y$ of

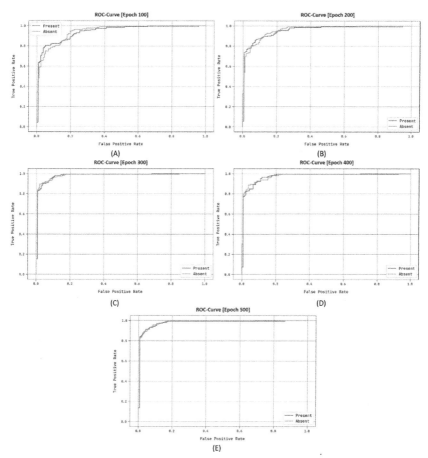

Figure 2.8 ROC of DEOFSOML-HDD system (A) Epoch100, (B) Epoch200, (C) Epoch300, (D) Epoch400, and (E) Epoch500.

88.76%, 85.43%, and 82.53%, respectively. Although the SVM model has demonstrated a considerable $accu_y$ of 92.20%, the DEOFSOML-HDD model has shown a superior $accu_y$ of 93.41%.

The comparative analysis of the DEOFSOML-HDD technique with existing ML techniques in $sens_y$ is exhibited in Fig. 2.10. The outcomes displayed that the DT and ANN techniques have attained reduced $sens_y$ of 88.43% and 59.35%, correspondingly. Then, the LOR, NB, and KNN techniques reached moderately closer $sens_y$ of 98.23%, 78.57%, and 74.10%, correspondingly. Although the SVM method has demonstrated a considerable $sens_y$ of 88.47%, the DEOFSOML-HDD method has displayed a superior $sens_y$ of 93.41%.

Table 2.3 Comparative analysis of the DEOFSOML-HDD system with recent approaches.

Methods	Accuracy	Sensitivity	Specificity	Computational time (seconds)
DEOFSOML-HDD	93.41	93.41	93.41	0.13
Support vector machine	92.20	88.47	98.02	5.70
LOR model	88.76	98.23	96.38	1.24
Naïve Bayes algorithm	85.43	78.57	88.67	2.13
KNN algorithm	82.53	74.10	87.28	3.28
Decision tree algorithm	79.67	88.43	91.72	0.92
ANN model	75.08	59.35	87.84	23.36

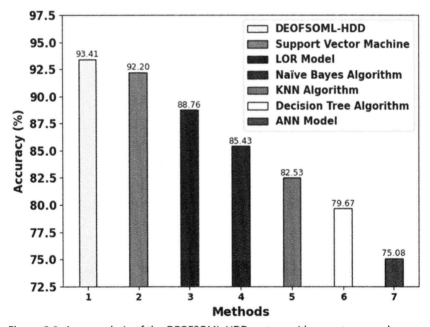

Figure 2.9 $Accu_y$ analysis of the DEOFSOML-HDD system with recent approaches.

The close inspection of the DEOFSOML–HDD technique's existing ML methods in $spec_y$ is shown in Fig. 2.11. The outcomes displayed that the DT and ANN models have obtained reduced $spec_y$ of 91.72% and 87.84%, correspondingly. Then, the LOR, NB, and KNN methods have achieved a moderately closer $spec_y$ of 96.38%, 88.67%, and 87.28%, respectively.

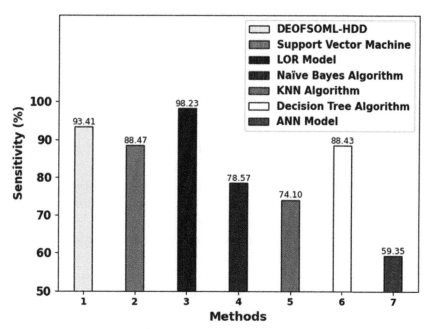

Figure 2.10 *Sens$_y$* analysis of the DEOFSOML-HDD system with recent methods.

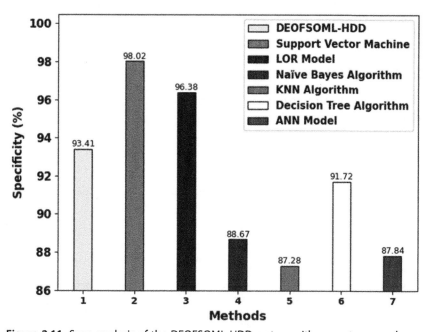

Figure 2.11 *Spec$_y$* analysis of the DEOFSOML-HDD system with current approaches.

Although the SVM approach has demonstrated a considerable *spec*$_y$ of 98.02%, the DEOFSOML-HDD technique has shown a superior *spec*$_y$ of 93.41%. The e results reassured the supremacy of the DEOFSOML-HDD model on HD classification.

2.5 Conclusion

In this study, we have introduced the DEOFSOML-HDD technique to improve HD detection efficiency in the IoMT environment. At the initial stage, the DEOFSOML-HDD technique uses IoT devices to gather patient data for disease detection. Data preprocessing transforms the raw medical data into a useful format. Besides, the DEOFS technique is exploited to optimally choose the subset of features. At last, the MBASA with VAE model is exploited for the disease detection process. An extensive set of simulations are carried out to examine the performance of the DEOFSOML-HDD approach. A detailed comparative study revealed the supremacy of the DEOFSOML-HDD technique over other existing models. Thus, the presented DEOFSOML-HDD technique can be applied to diagnose diseases in a real-time environment. In future, we plan to extend the performance of the EOFSOML-HDD technique by outlier removal and data clustering approaches.

Conflicts of interest

The authors declare that they have no conflicts of interest to report regarding the present study.

References

[1] Y.S. Su, T.J. Ding, M.Y. Chen, Deep learning methods in internet of medical things for valvular heart disease screening system, IEEE Internet of Things Journal 8 (23) (2021) 16921−16932.
[2] S. Tuli, N. Basumatary, S.S. Gill, M. Kahani, R.C. Arya, et al., HealthFog: an ensemble deep learning based smart healthcare system for automatic diagnosis of heart diseases in integrated IoT and fog computing environments, Future Generation Computer Systems 104 (2020) 187−200.
[3] C. Li, X. Hu, L. Zhang, The IoT-based heart disease monitoring system for pervasive healthcare service, Procedia computer science 112 (2017) 2328−2334.
[4] O. Deperlioglu, U. Kose, D. Gupta, A. Khanna, A.K. Sangaiah, Diagnosis of heart diseases by a secure internet of health things system based on autoencoder deep neural network, Computer Communications 162 (2020) 31−50.

[5] I.S. Brites, L.M. Silva, J.L. Barbosa, S.J. Rigo, S.D. Correia, *et al.*, Machine learning and iot applied to cardiovascular diseases identification through heart sounds: a literature review, in: International Conference on Information Technology & Systems, Lecture Notes in Networks and Systems book series, Springer, Cham, 2022, vol. 414, pp. 356−388.

[6] P.M. Kumar, U.D. Gandhi, A novel three-tier Internet of Things architecture with machine learning algorithm for early detection of heart diseases, Computers & Electrical Engineering 65 (2018) 222−235.

[7] S.S. Sarmah, An efficient IoT-based patient monitoring and heart disease prediction system using deep learning modified neural network, IEEE Access 8 (2020) 135784−135797.

[8] P. Verma, R. Tiwari, W.C. Hong, S. Upadhyay, Y.H. Yeh, FETCH: a deep learning-based fog computing and IoT integrated environment for healthcare monitoring and diagnosis, IEEE Access 10 (2022) 12548−12563.

[9] M. Aljanabi, M.H. Qutqut, M. Hijjawi, Machine learning classification techniques for heart disease prediction: a review, International Journal of Engineering & Technology 7 (4) (2018) 5373−5379.

[10] S. Yadav, V. Kadam, S. Jadhav, Machine learning algorithms for the diagnosis of cardiac arrhythmia in IoT environment, International Conference on Recent Trends in Image Processing and Pattern Recognition, Communications in Computer and Information Science book series, 1381, Springer, Singapore, 2021, pp. 95−107.

[11] C. Guo, J. Zhang, Y. Liu, Y. Xie, Z. Han, et al., Recursion enhanced random forest with an improved linear model (RERF-ILM) for heart disease detection on the Internet of Medical Things platform, IEEE Access 8 (2020) 59247−59256.

[12] A. Sheeba, S. Padmakala, C.A. Subasini, S.P. Karuppiah, MKELM: Mixed kernel extreme learning machine using BMDA optimization for web services based heart disease prediction in smart healthcare, Computer Methods in Biomechanics and Biomedical Engineering 25 (2022) 1−15.

[13] W. Lu, J. Jiang, L. Ma, H. Chen, H. Wu, et al., An arrhythmia classification algorithm using C-LSTM in physiological parameters monitoring system under internet of health things environment, Journal of Ambient Intelligence and Humanized Computing (2021) 1−11. Available from: https://doi.org/10.1007/s12652-021-03456-7.

[14] W. Ning, S. Li, D. Wei, L.Z. Guo, H. Chen, Automatic detection of congestive heart failure based on a hybrid deep learning algorithm in the internet of medical things, IEEE Internet of Things Journal 8 (16) (2020) 12550−12558.

[15] Q.U.A. Mastoi, T.Y. Wah, R.G. Raj, A. Lakhan, A novel cost-efficient framework for critical heartbeat task scheduling using the Internet of medical things in a fog cloud system, Sensors 20 (2) (2020) 441.

[16] M. Abdel-Basset, A. Gamal, G. Manogaran, L.H. Son, H.V. Long, A novel group decision making model based on neutrosophic sets for heart disease diagnosis, Multimedia Tools and Applications 79 (15) (2020) 9977−10002.

[17] G.T. Reddy, M. Reddy, K. Lakshmanna, D.S. Rajput, R. Kaluri, et al., Hybrid genetic algorithm and a fuzzy logic classifier for heart disease diagnosis, Evolutionary Intelligence 13 (2) (2020) 185−196.

[18] M.A. Khan, F. Algarni, A healthcare monitoring system for the diagnosis of heart disease in the IoMT cloud environment using MSSO-ANFIS, IEEE Access 8 (2020) 122259−122269.

[19] J.S. Kallimani, R. Walia, B. Belete, A novel feature selection with hybrid deep learning based heart disease detection and classification in the e-healthcare environment, Computational Intelligence and Neuroscience 2022 (2022) 1−12.

[20] A. Faramarzi, M. Heidarinejad, B. Stephens, S. Mirjalili, Equilibrium optimizer: A novel optimization algorithm, Knowledge-Based Systems 191 (2020) 105190.

[21] A. González-Muñiz, I. Díaz, A.A. Cuadrado, D. García-Pérez, D. Pérez, Two-step residual-error based approach for anomaly detection in engineering systems using variational autoencoders, Computers and Electrical Engineering 101 (2022) 108065.

[22] S.W. Fei, Y.Z. Liu, Fault diagnosis method of bearing utilizing GLCM and MBASA-based VAE, Scientific Reports 12 (1) (2022) 1−8.

CHAPTER 3

Artificial intelligence—based security attack detection for healthcare cyber-physical system: lightweight deep stochastic learning

D. Prabakar[1], Shamimul Qamar[2] and R. Manikandan[3]
[1]Department of Data Science and Business Systems, School of Computing, Faculty of Engineering and Technology, SRM Institute of Science and Technology, Chengalpattu, Tamil Nadu, India
[2]Computer Science & Engineering Department, College of Sciences & Arts, Dhahran Al Janoub Campus, King Khalid University, Abha, Kingdom of Saudi Arabia
[3]School of Computing, SASTRA Deemed University, Thanjavur, Tamil Nadu, India

3.1 Introduction

Internet of Things (IoT) is widely regarded as one of the most exciting technologies of the information technology age. Every day, a number of connected devices grow exponentially using Internet, and by 2020, more than 50 million devices will be connected to the Internet [1]. IoT technology aims to connect all objects in such a way that they become programmable, intelligent computers and make human communication safer. Everything has direct access to sensors and networks, allowing for the exchange of vital information. In the future, machine-to-machine communication is feasible. IoT has a plethora of practical applications that can be applied almost everywhere, including smart city applications (smart home, smart grid, healthcare, and other areas) where these applications improve quality of life [2]. The intrusion detection system (IDS) is designed to stop incidents and detect threats or intrusions into the network. It also actively monitors the network by spotting potential events and recording information about them. Computational, networking, and physical processes are just a few of the many entities that are integrated in cyber—physical systems (CPSs). In general, networked medical CPS has the potential to provide a number of advantages, including the ability to carry out monitoring and treatment procedures in a manner that is both

Securing Next-Generation Connected Healthcare Systems
DOI: https://doi.org/10.1016/B978-0-443-13951-2.00009-X

more effective and less costly. Current medical device systems are an exceptional CPS example because they integrate embedded applications, networking capabilities, and the intricate physical dynamics of the human body. One report assessed that organized clinical CPS advancements can set aside to 63 billion in medical services costs for the accompanying 15 years, lessening 15%—30% with respect to clinical hardware costs [3]. By ensuring safe, personalized treatment, medical CPS aims to improve patient care's efficiency and effectiveness. Medical CPS differs from conventional network environments in two important ways [4]. To begin, the information that is sent between medical devices is extremely private and highly sensitive for both patients and medical organizations. Because of this, cybercriminals can use this sensitive data to their advantage and make a profit. Second, a large number as well as variety of medical devices, particularly those that are connected to the Internet and may be susceptible to a wider range of cyber threats make the infrastructure of medical CPS frequently challenging. For instance, healthcare providers reported a 60% increase in number of data security breaches between 2013 and 2014, compared to a 30% increase in other industries [5]. The security of medical CPS ought to receive significantly more attention in light of the rapid change in the medical industry. Utilizing machine learning (ML) requires evaluating and modeling new network behavior in terms of trustworthiness. Methods for ML take into account the best model for dealing with the massive and never-ending volume of data provided by IoT devices. A technique for disseminating interpretations and predictions based on in-depth analysis of data patterns is called deep learning (DL). IoT conditions utilize different DL procedures to recognize examples of strange way of behaving. We are able to model a prediction system that is both effective as well as adaptable. We can do this by utilizing DL techniques. Assessment of false-positive (FP) and false-negative rates during prediction as well as examination of features' misperception-related aspects during learning are both areas of the study that are lacking [6].

3.2 Related works

IoT has a significant impact on growth of the healthcare sector. Additionally, as physical devices connect to the Internet to share real-time data with the medical team and collect vital signs with a variety of sensors, it has emerged as a significant and significant source of medical data [7]. A system that uses embedded wearable sensors to remotely monitor health specifications, stores analyzed data in cloud, and automatically sends analysis results to a doctor when a condition is critical was proposed in work [8]. By minimizing number of visits to doctor, the proposed system reduces healthcare costs [9]. The author [10] developed and implemented a health monitoring system for regulating human health parameters in a safe motherhood program. A robust IDS that makes use of IoT for research as well as makes use of CNN with LSTM and PSO for feature selection is described in the work by [11]. An IDS was developed by author [12] by combining the fuzzy MutiVariateOptimizer algorithm with ANN to identify distinct threats using the NSL-KDD and UNSW-NB15 datasets. To identify various IoT threats, work [13] conducted data analysis on the public datasets NSL-KDD, DS2Os, IoTDevNet, IoTID20, and IoTBotnet for an IDS in the IoT environment. Using ensemble algorithms and data produced for IP cameras and IoT devices using the MILSTD-1553 communication protocol for hierarchical data, work [14] developed methods to improve detection performance and reduce FP rates. Using fuzzy rough set-based feature selection as well as genetic method-based learning strategy, Ref. [15] built an adaptive IDS with higher level accuracy. An intelligent approach to the creation of deep neural network-based IDS was developed in work [16] by maximizing fitness value hashing to locate abnormalities in networks utilizing CICIDS2017 dataset. Based on DL, the intelligent IDS developed by [17] detects unpredictable cyberattacks. The datasets NSLKDD, UNSW-NB15, Kyoto, WSN-DS, and CICIDS17 were used to train and test the framework during the research. Using both supervised and unsupervised feature selection algorithms for intrusion detection, work [18] presented a novel two-layer dimension reduction model for the purpose of detecting low-frequency attacks. Using a variety of protocols, this model is also suggested for detecting attacks not only at the network layer but also at other layers like the application and support layers. Ref. [19] created a model to identify unexpected threats by combining binary classification and a classification algorithm for deep neural networks. Without knowing the payload of the packet, this detection is carried out using connection parameters and network packet analysis. Work [20] introduced a novel data reduction technique using K-means $+ +$ clustering and the SVM-SMOTE oversampling technique with SLFN Classification to reduce processing value of computation by reducing number of instances in IoTID20 dataset. IDS involves monitoring and looking for signals of possible events that violate information security, information protection laws, and customary security practices in a computer network that has been configured.

3.3 System model

In this section, ML techniques have been used to improve the novel technique based on cyber-security analysis and CPS. Lightweight deep stochastic learning-based cyber-attack detection improves network security. The healthcare network is then built using quantum cloud federated learning (FL) on top of a CPS.

T_e natural information were acquired. Eventually, various techniques were utilized to eliminate copies and invalid qualities, among them. This method is utilized in data mining to convert raw data into a format that are understood. However, data from the real world may contain inconsistencies, gaps, or both in some instances. The following steps are taken before processing:

Predictive modeling is hindered by skewed classification. The majority of categorization ML methods have same number of instances for each class. Methods from minority groups are therefore underrepresented. This raises a red flag when you consider that minorities are more likely than dominant group to be misclassified. Research dataset has been cleaned up by getting rid of any outliers. The method of resampling has changed significantly as a result of these studies. For instance, collecting records from every cluster under sampling can aid in data preservation. Oversampling can be used to create synthetic samples that are more diverse than exact duplicates of minority class data.

3.3.1 Lightweight deep convolutional stochastic learning−based cyber-attack detection

Given a traffic trace file from dataset and a time frame with a predetermined duration of t seconds, the algorithm extracts 11 properties from each packet with a capture time between t0, capture time of first packet, and time $t_0 + t$. We automatically eliminate attributes that would prevent the model from being generalized, such as link layer encapsulation type and application-layer attributes, which are particular to end-hosts and user applications. Similar to how online IDSs capture traffic, which is then delivered to algorithms for anomaly detection after being captured for a predetermined period of time (t). As a result, these algorithms are forced to make judgments based on segments of traffic flows without being fully aware of their complete existence.

To automatically collect malware, the proposed model in Fig. 3.1 will make use of a Virtual Honeypot with low and medium interactions, respectively. The operating system is only partially accessible through a low-interaction honeypot. It is not intended to be a fully featured operating system by design, and it typically cannot be fully exploited. As a result, 0-day exploits are not well captured by low-interaction honeypots. Instead, it can be used to figure out how often your network is attacked and find known exploits. There are numerous benefits to low-interaction honeypots.

Our goal is to make this CNN model as simple and quick to run as possible so that it can be used on devices with limited resources. To accomplish this, proposed method is a lightweight, regulated location framework that integrates a CNN, like that from the field of Regular Language Handling. When it comes to the weights of the kernels, CNNs use parameters that are shared and reused, whereas a conventional neural network only makes use of each weight once. Our model requires less storage and memory as a result.

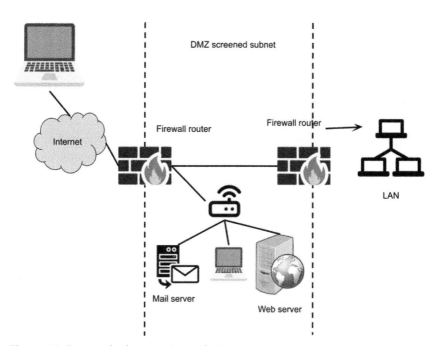

Figure 3.1 Proposed cyber-security analysis.

Layer of CNN: (Figure 1.2), a single convolutional layer employs k filters of size h_f on each input matrix F, where h is length of each filter and $f = 11$ once more. In accordance with CNN convention, we employ $\text{ReLU}(x) = \max 0$, x to introduce nonlinearity in the learned filters. An activation matrix A of size $(n - h + 1)$ k is created when all activation maps are stacked, with $A = [a_1 | \ldots | a_k]$. The incorporation of a CNN into our architecture has two primary advantages.

Second, the model can achieve efficiency improvements over traditional NN since weights of each filter are reused across whole input. When compared to a standard neural net, which relies on full end-to-end connectivity, sharing weights reduces number of learnable parameters and makes method lighter. Second, CNN learns weights as well as biases of every filter automatically during training phase, encapsulating learning of important characteristics and features in final method during training. This method is more adaptive to new DDoS attack variations since training phase may be quickly rerun whenever new training data are available without the need to develop as well as rank new features.

Pooling layer maximum: For maximum pooling, we downsample along A's first dimension, which represents temporal character of input. The greatest m activations of each learnt filter are contained in an output matrix m_o of size $((n \times h + 1)/m)$ k, so that $m_o = [\max(a_1) | \ldots | \max(a_k)]$ when pool size is m. The model thus prioritizes the larger activations over the less useful information that resulted in smaller activations. This also means that we get rid of activation's positional data, which tells us where it happened in the original flow. This makes the feature encoding smaller and makes the network simpler. Final one-dimensional feature vector, v, which will be fed into classification layer, is created by flattening mo.

Layer of classification: The output layer has a single node and is connected to a fully connected layer of same size as input layer. Sigmoid activation function receives output x in form "$(x) = 1/(1 + e \times x)$." The probability that a given flow is a malicious DDoS attack is returned as $p[0,1]$ because the activation is limited to a value between 0 and 1. If p is greater than 0.5, the flow is categorized as DDoS, while it is benign otherwise.

3.3.2 Cyber–physical system using quantum cloud federated learning in healthcare system

QL is a type of reinforcement learning that chooses the best behavior based on experience after using a trial-and-error method to investigate complex as well as stochastic environment. Concepts of state, action, and reward all exist in QL. Environment-specific conditions are seen as state, behavior as the action, and experience as the reward. According to Eq. (3.1), an action ( ) performed in a state (s_t) advances to the following state ($s_t + 1$).

$$s_t \xrightarrow{a_t} s_{t+1} \tag{3.1}$$

$$\mathscr{Q}(s_t, a_t) \leftarrow \mathscr{Q}(s_t, a_t) + \eta \cdot (r_{t+1} + \gamma \cdot \max_a \mathscr{Q}(s_{t+1}, a_{t+1}) - \mathscr{Q}(s_t, a_t)) \tag{3.2}$$

This model illustrates the point of decision-making for mold scheduling by the following equations:

$$\min \sum_i w_i T_i$$

$$T_i \geq C_i - d_i, \forall i \in I$$

$$st_{ipv} \geq st_{ip(o-1)} + \sum_k pt_{ip(o-1)} \cdot x_{ip(o-1)k}, \forall i \in I, p \in \mathscr{P}_i, o \in O_p^i, o > 1, k \in \mathscr{M}_{p(0-1)}^i$$

$$st_{ipu} \geq st_{ippoo} + pt_{ippoor} + su_{ipouppoork} - (2 - x_{ipak} - x_{ippoork} + y_{ipoippor})H, \forall i, ir \in T \tag{3.3}$$

$$p \in \mathscr{P}_i, p' \in \mathscr{P}_i, k \in \mathscr{M}_p^i \cap \mathscr{M}_{p,'}^{i'}, o \in O_p^i, or \in O_{p''}^i \mid o \neq \text{ or}$$

$$st_{i\ ippor} \geq st_{ipo} + pt_{ipo} + su_{ipoipprork} - (3 - x_{ipok} - x_{ippork} + y_{ipoippor}) \cdot H, \quad Vi, \ ir \in I$$

$$p \in \mathscr{P}_i, p' \in \mathscr{P}_i, k \in \mathscr{M}_p^i \cap \mathscr{M}_p^i, o \in O_p^i, or \in \mathcal{O}_{p''}^{''} \mid o \neq o'$$

$$st_{ipo} \geq r_i, \forall i \in I, p \in \mathscr{P}_i, o \in O_p^j, o = 1$$

$$C_{ip} \geq st_{ipo} + \sum_k pt_{ipo} \cdot x_{ipok}, \forall i \in I, p \in \mathscr{P}_i, \in O_{p,}^j o \in \left| O_p^i \right|, k \in \mathscr{M}_p^i \tag{3.4}$$

$$C_i = \max_{p \in P_i} C_{ip}, \forall i \in I$$

$$\sum_k x_{ipok} = 1, \forall i \in I, p \in \mathscr{P}_i, o \in O_p^i k \in \mathscr{M}_p^i T_i, C_i, st_{ipo}, C_{ip} \geq 0, x_{ipok}, y_{ijpp'}{}'_{p'} \in \{0, 1\}$$

The MDP, which can mathematically method decision-making process, is suggested in this section for deep RL. Through transition probability, state shifts to next state, $s_t + 1$, and the environment provides the reward, r_t. At each distinct moment, the same procedure is carried out. State, action, and reward are discussed in following section, but the transition probability is not included because deep RL may or may not require it.

The job and the machine make up the majority of the information in the environment of the mold production scheduling system. In this system, state is meant to be a collection of idle machine statuses. Assume that K_t is a collection of idle machines, and that ty_k^t stands for machine setup type for K_t at time t. Machines are assigned to specific operation kinds, $sy = $ "1, 2,..., SY," and additional setup time is required if the machine is given an additional operation type. Let SO_t^{SY} be the total of the operations with operation type sy in O_t^k, and O_t^k be a collection of waiting operations for machine k at time t. The work has weight $jw = 1, 2,...,$ JW, as stated in Section 2.2, and let OW_t^{jw} be total of operations that have jw at time t. State of each machine k at time t, s_t^k, is stated as Eq.(3.5) based on the foregoing explanation.

$$s_t^k = \left\{ ty_k^t, \left(\frac{OW_t^1}{|\overline{O}_t^k|}, \frac{OW_t^2}{|\overline{O}_t^k|}, \cdots \frac{OW_t^{JW}}{|\overline{O}_t^k|} \right), \left(\frac{SO_t^1}{|\overline{O}_t^k|}, \frac{SO_t^2}{|\overline{O}_t^k|}, \cdots \frac{SO_t^{SY}}{|\overline{O}_t^k|} \right) \right\}, \square, k, t$$

(3.5)

The state vector in Eq. (3.5), which is given as a collection of vector numbers with values between [0,1], is standardized by dividing total number of operations, O_t^k. Agent can travel to all the states by doing this. Last but not least, s_t can be written as follows. $s_t = \left\{ s_t^1, s_t^2, \ldots, s_t^K \right\}, \forall t$

The choice of operation to assign an idle machine k is referred to as the action. As a result, total number of waiting operations for machine k is number of candidates for action, which is action space. Action space for s_t^k is shown in Eq. (3.6).

$$a_t^k \in \left\{ \overline{O}_t^k \right\}, \forall k, t$$

(3.6)

Eq. (3.7) enables the following description.

$$a_t = \left\{ a_t^1, a_t^2, \ldots, a_t^K \right\}, \quad \forall t$$

(3.7)

Layered architectures of our proposed system as well as those proposed are compared in Fig. 3.2. The proposed framework design is made out of an actual layer, network layer, IoT blockchain cloud layer, application layer, business layer. ML algorithm is not used in the proposed system, which uses blockchain-based smart contracts to monitor the vital signs of patients. The security concerns surrounding intrusion detection were not discussed by the system. Because it is centralized, the system has a single point of failure and sends medical data to a centralized server, which can compromise the data.

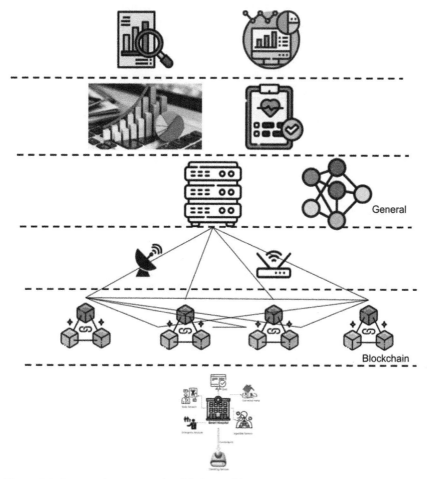

Figure 3.2 Proposed quantum cloud federated learning.

Layer for perception of healthcare data: this includes sensor-equipped ICU IoT devices. There are two types of case devices for the ICU: devices for monitoring the health and environment of the room and the patients.

Layer of edge-based blockchain: IoT gateways make up this. There are some healthcare sensing devices in each gateway. Physical healthcare sensors do not have a global internal protocol; As a result, the gateways were able to support a wide range of network access protocols. The detection of multiple attacks is the job of an IoT gateway. It will do this by preventing data from its gateway. The closer the attack resources are, the shorter the intrusion's detection time will be. Because FL method works with smaller data sets, there will also be less computing and processing power available. A further application for these chained blocks will be aggregation as well as averaging. Lastly, that chain is secured by a cryptographic hash function that links blocks together in chain. As a result, it cannot be altered or manipulated because consensus algorithms (smart contracts) govern its operation. After that, ES combines the ciphertext and sends it along with its signature to activated blockchain layer, which is managed by a smart contract. This ensures that data integrity and privacy are maintained. Under the rules of the smart contract, when a smart contract receives the data of all ESs, it verifies validity of these messages by utilizing public keys of ESs and stored data as hashed blocks on blockchain. Using its own secret key, CPCC can retrieve aggregated plaintext and stored blocks from activated blockchain. Through blockchain network and ES, the local method can upload and download data to or from the CPCC, just as it can do with data storage that is based on terminal edge computing.

Layer on top: this is liable for getting exchange of information from lower layer to higher layer. It is referred to as connectivity layer because its function is to manage routing.

Layer of application: This is in charge of keeping an eye on vital signs in healthcare.

Layer of business: This enables healthcare application service managers to produce business models, flowcharts, and executive reports based on the examined and acquired data from lower layers.

By cooperatively learning a model without sharing personal data between users, FL aims to safeguard data privacy. Existing privacy-preserving solutions are insufficient for Federated Learning of Deep Neural Networks with billions of method parameters. Methods based on homomorphic encryption provide secure privacy protections, but they have extremely high overheads for computation and communication, making them almost useless in practice. Profound learning with differential protection was executed as a useful learning calculation at a reasonable expense in intricacy. We take into account a federated DL scenario in which K participants learn a multilayered deep neural network model together without disclosing their private training data.

We assume that some participants or the server may be a semihonest adversary because they may launch privacy attacks on information exchanged by other participants but do not send any erroneous messages. A single client has a private data $x \in X$ and exposed public information $G \in \Omega$ which is related to private data by a mapping $G = g(x)$, $X \to \Omega$. Note that the mapping $g()$ is assumed to be known to all participants in FL, while data x is kept secret only to its owner. In case that $g()$ is an invertable function, it is straightforward for adversaries to infer data $x = g - 1$ (G) from the exposed information. In general $g()$ is NOT invertable, for instance, in the context of federated deep neural network learning, $g(x)$ is the exchanged neural network parameters or its gradients depending on input training data x. For classification tasks, the NN model parameters G is sought by minimizing a loss function defined on training data x by the following equation:

$$\mathscr{G}^* = \min_{\mathscr{G}} L_{CE}(\mathscr{G}, x) \tag{3.8}$$

where $L()$ is often cross-entropy (CE) loss for the classification task. Even $g()$ in (1) is NOT invertable, nevertheless, adversaries may still use $\tilde{x}$ to estimate x in a Bayesian restoration sense by minimizing $|g(\tilde{x}) - G|$.

Attack mechanism with Bayesian federated learning:

A privacy leakage attack is an optimization process A that aims to restore the original data such that the restored data $\tilde{x}$ best fits the exposed information $\tilde{\mathscr{G}}$, that is, by Eq. (3.9).

$$\tilde{x}^* = \mathscr{A}(\tilde{\mathscr{G}}) := \arg\min_{\hat{x}} L(g(\tilde{x}), \tilde{\mathscr{G}}) \tag{3.9}$$

where $L(g(\tilde{x}), \tilde{\mathscr{G}})$ is a loss function depending on specific attacking methods adopted by the attacker. The attacking in Eq. (3.10) is derived from the Bayesian data restoration framework, which dictates that inference about the true data x is governed by

$$P(\bar{x} \mid \tilde{\mathscr{G}}) = \frac{P(\tilde{\mathscr{G}} \mid \bar{x}) P(\bar{x})}{P(\tilde{\mathscr{G}})} \propto P(\tilde{\mathscr{G}} \mid \tilde{x}) P(\tilde{x}) \tag{3.10}$$

where $P(\tilde{x}\,|\,\tilde{\mathscr{G}})$ is the probability of $\tilde{x}$ conditioning on the observed information $\tilde{\mathscr{G}}$, and $P(G\,|\,\tilde{x})$ is the likelihood of observed information. Thus, the best restored data $\tilde{x}^*$ is sought by the optimization process by the following equation:

$$\tilde{x}^* = \mathscr{A}\left(\overline{\mathscr{G}}\right) = \arg\max_{\tilde{x}} \log\left(P\left(\overline{x}\,|\,\overline{\mathscr{G}}\right)\right) = \arg\min_{\tilde{x}}[-\log\left(P\left(\overline{\mathscr{G}}\,|\,\overline{x}\right)\right)$$

$$-\log\left(P\left(\overline{x}\right)\right)] \triangleq \arg\min_{\tilde{x}}\left[L_1(g(\tilde{x}),\tilde{\mathscr{G}}) + L_2(\tilde{x})\right]$$

$$(3.11)$$

where the minimization of $L2 = -\log((P(\tilde{x})))$ enforces a prior of x, such as image smoothness prior by Total Variation loss T V $(\tilde{x})$. As for the negative log likelihood (NLL) loss $L1 = -\log(P(G\,|\,\tilde{x}))$.

3.4 Performance analysis

The investigations were performed on a Linux arrangement with 16 GB Slam and Intel Center i7 10-th Age Processor. Using TensorFlow/Keras/Scikit-learn stack and the Python programming language, all of the experiments were carried out.

Description of dataset: Labeled datasets of network traffic make it easier for supervised methods to provide the necessary data for effective IDS training to achieve remarkable accuracy and dependability in distinguishing a variety of network attacks. IoT-23, LITNET-2020, and NetML-2020 are just a few of the new flow-based benchmark datasets that have recently been made available to the general public. These have not yet been widely utilized by cyber-security community due to their temporal proximity. On the other hand, we make use of these datasets for effective anomaly-based network intrusion detection based on real-time network traffic data.

IoT-23 is a network traffic dataset with 20 malware subsets and three benign subsets. The Stratosphere Laboratory in Czechia made the dataset available for the first time in January 2020. The dataset's goal is to provide ML-based intrusion detection tools with a substantial set of labeled malware and benign traffic from real captures. In total, the 20 malicious captures bear the following labels: DDoS, C&C-Heart Beat, C&C, Attack, C&C, C&C-Heart Beat Attack, C&C-File download, C&C-Tori, File download, C&C-Heart Beat File Download, Part-Of-A-Horizontal-PortScan On the other hand, 30,858,735 flows belong to the benign category. Nevertheless, the class label is one of 21 feature attributes in the dataset. As a result, there are a total of 21 attributes included in each data instance, which define the properties of connections. The qualities range from being nominal to numerical to having time-stamped values.

Senders and collectors make up LITNET-2020 NetFlow dataset. The senders, which were used to calculate NetFlow data that was passing through collectors, are made up of Cisco routers and Fortige (FG-1500D) firewalls. Software that handles data receiving, storing, and filtering is included in the collector. Ordinary data and malicious data are two types of instances. Taking into account kind of network attack, malicious instances are further divided into nine categories.

NetML-2020: The 30 traffic data from Stratosphere IPS were utilized to create NetML-2020 dataset, which was used for anomaly detection tasks. By providing a raw pcap file as an input to feature extraction tool, flow features are extracted in JavaScript Object Notation format, and each flow sample is listed in output file. Finally, a one-of-a-kind id number is assigned to each flow to identify the label information and raw traffic packet capture file. Only 26 meta-features were chosen after data preprocessing stage since we only care about "top-level" granularity, despite the fact that the NetML dataset contains 48 feature attributes and 48 flows.

3.4.1 Proposed analysis

Table 3.1 shows parametric analysis based on proposed technique for various secure dataset. The dataset analyzed are IoT-23, LITNET-2020, and NetML-2020 in terms of network efficiency, security analysis, training accuracy, validation loss, f-measure.

Fig. 3.3 shows parametric analysis of IoT-23 dataset in terms of network efficiency of 93%, security analysis of 91%, training accuracy of 81%, validation loss of 55%, f-measure of 85% based on proposed technique.

Fig. 3.4 shows parametric analysis for LITNET-2020 dataset for proposed techniques. The proposed technique attained network efficiency of 92%, security analysis of 89%, training accuracy of 79%, validation loss of 49%, f-measure of 83%.

Table 3.1 Parametric analysis based on proposed techniques.

Dataset	Network efficiency	Security analysis	Training accuracy	Validation loss	f-Measure
IoT-23	89	88	77	45	79
LITNET-2020	92	89	79	49	83
NetML-2020	93	91	81	55	85

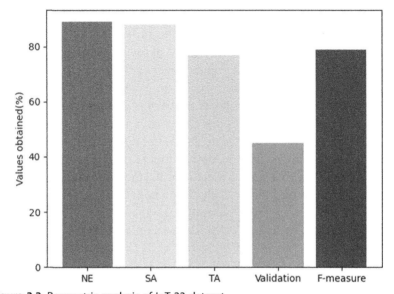

Figure 3.3 Parametric analysis of IoT-23 dataset.

Fig. 3.5 shows parametric analysis of NetML-2020 dataset in terms of network efficiency of 89%, security analysis of 88%, training accuracy of 77%, validation loss of 45%, f-measure of 79% based on proposed technique (Table 3.2).

3.4.2 Comparative analysis

Fig. 3.6 shows comparison of network efficiency between proposed and existing technique for IoT-23, LITNET-2020, and NetML-2020. Proposed technique attained network efficiency of 89%, while existing MILSTD-1553 attained network efficiency of 81%, SVM_SMOTE attained network efficiency of 85% for a IoT-23 dataset; for LITNET-2020 dataset the proposed technique attained network efficiency of 92%, while existing MILSTD-1553 attained network efficiency of 90%,

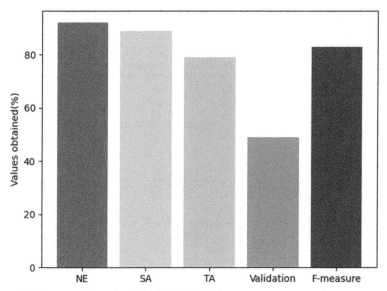

Figure 3.4 Parametric analysis of LITNET-2020 dataset.

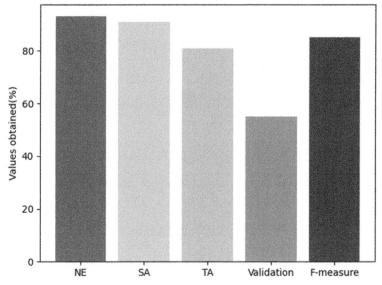

Figure 3.5 Parametric analysis of NetML-2020 dataset.

SVM_SMOTE attained network efficiency of 91%, the proposed technique attained network efficiency of 93%, while existing MILSTD-1553 attained network efficiency of 91%, SVM_SMOTE attained network efficiency of 92% for a NetML-2020 dataset.

Table 3.2 Comparative analysis between proposed and existing technique based on various security attack dataset.

Dataset	Techniques	Network efficiency	Security analysis	Training accuracy	f-Measure
IoT-23	MILSTD-1553	81	84	74	71
	SVM_SMOTE	85	86	75	75
	AI_SAD_LDSL	89	88	77	79
LITNET-2020	MILSTD-1553	90	86	71	81
	SVM_SMOTE	91	88	75	82
	AI_SAD_LDSL	92	89	79	83
NetML-2020	MILSTD-1553	91	88	77	82
	SVM_SMOTE	92	90	80	83
	AI_SAD_LDSL	93	91	81	85

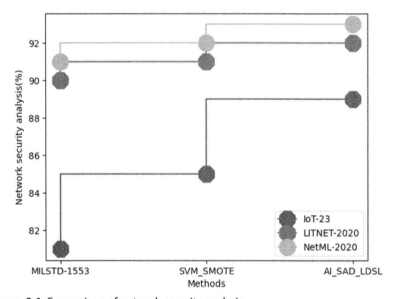

Figure 3.6 Comparison of network security analysis.

Fig. 3.7 shows comparison of Security analysis between proposed and existing techniques. Proposed technique attained Security analysis of 88%, while existing MILSTD-1553 attained Security analysis of 84%, SVM_SMOTE attained Security analysis of 86% for a IoT-23 dataset; for LITNET-2020 dataset proposed technique attained Security analysis of 89%, while existing MILSTD-1553 attained Security analysis of 89%, SVM_SMOTE attained Security analysis of 86%, proposed technique

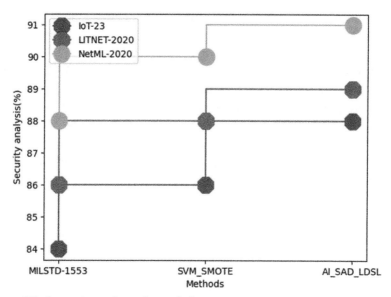

Figure 3.7 Comparison of security analysis.

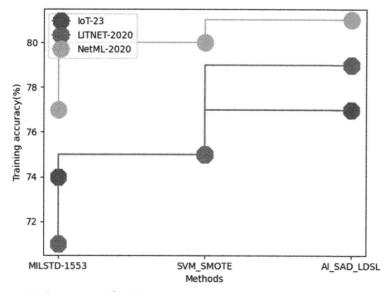

Figure 3.8 Comparison of training accuracy analysis.

attained Security analysis of 91%, while existing MILSTD–1553 attained Security analysis of 89%, SVM_SMOTE attained Security analysis of 90% for a NetML–2020 dataset.

Fig. 3.8 shows comparison of training accuracy between proposed and existing technique for IoT–23, LITNET–2020, and NetML–2020. Proposed technique attained Training accuracy of 77%, while existing MILSTD-1553 attained Training accuracy of 74%, SVM_SMOTE attained training accuracy of 75% for a IoT–23 dataset; for LITNET–2020 dataset the proposed technique attained training accuracy of 79%, while existing MILSTD-1553 attained training accuracy of 71%, SVM_SMOTE attained Training accuracy of 75%, the proposed technique attained training accuracy of 81%, while existing MILSTD-1553 attained training accuracy of 77%, SVM_SMOTE attained training accuracy of 80% for a NetML–2020 dataset.

Fig. 3.9 shows the comparison of *f*-measure between proposed and existing techniques. The proposed technique attained *f*-measure of 79%, while existing MILSTD-1553 attained *f*-measure of 71%, SVM_SMOTE attained *f*-measure of 75% for a IoT–23 dataset; for LITNET–2020 dataset the proposed technique attained *f*-measure of 83%, while existing MILSTD-1553 attained *f*-measure of 81%, SVM_SMOTE attained *f*-measure of 82%, the proposed technique attained *f*-measure of 85%, while existing MILSTD-1553 attained *f*-measure of 82%, SVM_SMOTE attained *f*-measure of 83% for a NetML–2020 dataset.

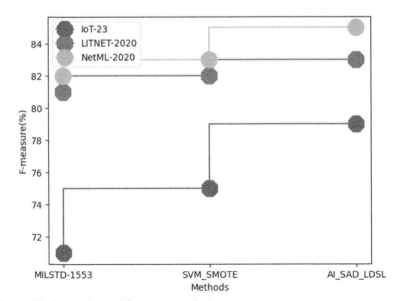

Figure 3.9 Comparison of *f*-measure analysis.

3.5 Conclusion

Based on CPS and DL techniques, this research proposes a novel method for healthcare system cyber-security analysis. Lightweight deep stochastic learning-based cyber-attack detection is utilized to enhance security of network, and quantum cloud FL is used to build a healthcare network based on a CPS. Cyber-attack behavior can be predicted by ML models, and processing this data can help healthcare professionals make decisions. This proposed method depends on a patient-driven plan that protects data on a confided in gadget. A secure DL operation is made possible by the blockchain, which takes control away from a centralized authority. The results of the experiment show that the proposed strategy encourages higher accuracy. The proposed technique attained network efficiency of 92%, security analysis of 89%, training accuracy of 79%, validation loss of 49%, *f*-measure of 83%.

References

[1] D. Hermawan, N.M.D.K. Putri, L. Kartanto, Cyber physical system based smart healthcare system with federated deep learning architectures with data analytics, International Journal of Communication Networks and Information Security 14 (2) (2022) 222–233.

[2] V. Ravi, T.D. Pham, M. Alazab, Attention-based multidimensional deep learning approach for cross-architecture IoMT malware detection and classification in healthcare cyber-physical systems, IEEE Transactions on Computational Social Systems 10 (4) (2023) 1597–1606.

[3] M. Abdullahi, H. Alhussian, N. Aziz, Adaptation of machine learning and blockchain technology in cyber-physical system applications: a concept paper, in: International Conference on Artificial Intelligence for Smart Community: AISC 2020, 17–18 December, UniversitiTeknologi Petronas, Malaysia, Springer Nature Singapore, Singapore, 2022, pp. 517–523.

[4] R. Verma, Smart city healthcare cyber physical system: characteristics, technologies and challenges, Wireless Personal Communications 122 (2) (2022) 1413–1433.

[5] R. Ch, G. Srivastava, Y.L.V. Nagasree, A. Ponugumati, S. Ramachandran, Robust cyber-physical system enabled smart healthcare unit using blockchain technology, Electronics 11 (19) (2022) 3070.

[6] P. Semwal, A. Handa, Cyber-attack detection in cyber-physical systems using supervised machine learning, Handbook of Big Data Analytics and Forensics, Springer, 2022, pp. 131–140.

[7] A. Alzahrani, M. Alshehri, R. AlGhamdi, S.K. Sharma, Improved wireless medical cyber-physical system (IWMCPS) based on machine learning, Healthcare, Vol. 11, MDPI, 2023, p. 384. No. 3.

[8] Z. Wang, Z. Li, D. He, S. Chan, A lightweight approach for network intrusion detection in industrial cyber-physical systems based on knowledge distillation and deep metric learning, Expert Systems with Applications 206 (2022) 117671.

[9] S.M. Nagarajan, G.G. Deverajan, A.K. Bashir, R.P. Mahapatra, M.S. Al-Numay, IADF-CPS: intelligent anomaly detection framework towards cyber physical systems, Computer Communications 188 (2022) 81−89.

[10] A.S. Rajawat, P. Bedi, S.B. Goyal, R.N. Shaw, A. Ghosh, Reliability analysis in cyber-physical system using deep learning for smart cities industrial IoT network node, AI and IoT for Smart City Applications, Springer, 2022, pp. 157−169.

[11] J. Singh, M. Wazid, A.K. Das, V. Chamola, M. Guizani, Machine learning security attacks and defense approaches for emerging cyber physical applications: a comprehensive survey, Computer Communications 192 (2022) 316−331.

[12] Z. Shen, F. Ding, A. Jolfaei, K. Yadav, S. Vashisht, K. Yu, DeformableGAN: generating medical images with improved integrity for healthcare cyber physical systems, IEEE Transactions on Network Science and Engineering 10 (2023) 2584−2596.

[13] R. Almajed, A. Ibrahim, A.Z. Abualkishik, N. Mourad, F.A. Almansour, Using machine learning algorithm for detection of cyber-attacks in cyber physical systems, Periodicals of Engineering and Natural Sciences 10 (3) (2022) 261−275.

[14] I. Priyadarshini, R. Sharma, D. Bhatt, M. Al-Numay, Human activity recognition in cyber-physical systems using optimized machine learning techniques, Cluster Computing 26 (2023) 2199−2215.

[15] M. Umer, S. Sadiq, H. Karamti, R.M. Alhebshi, K. Alnowaiser, A.A. Eshmawi, et al., Deep learning-based intrusion detection methods in cyber-physical systems: challenges and future trends, Electronics 11 (20) (2022) 3326.

[16] H. KeshmiriNeghab, M. Jamshidi, H. KeshmiriNeghab, Digital twin of a magnetic medical microrobot with stochastic model predictive controller boosted by machine learning in cyber-physical healthcare systems, Information 13 (7) (2022) 321.

[17] R.M. Richard, J.V. Taylar, Cyber-physical system framework for cerebrovascular accidents using machine learning algorithm, in: 2022 International Conference on ICT for Smart Society (ICISS), IEEE, August 2022, pp. 01−08.

[18] L.K. Ramasamy, F. Khan, M. Shah, B.V.V.S. Prasad, C. Iwendi, C. Biamba, Secure smart wearable computing through artificial intelligence-enabled Internet of Things and cyber-physical systems for health monitoring, Sensors 22 (3) (2022) 1076.

[19] M. Adil, M.K. Khan, M.M. Jadoon, M. Attique, H. Song, A. Farouk, An AI-enabled hybrid lightweight Authentication scheme for intelligent IoMT based cyber-physical systems, IEEE Transactions on Network Science and Engineering 10 (2022) 2719−2730.

[20] P. Kanagala, Effective cyber security system to secure optical data based on deep learning approach for healthcare application, Optik 272 (2023) 170315.

CHAPTER 4

Preparedness and impact of cyber secure system in clinical domain

Suvra Mukherjee, Ayush Pal and Sushruta Mishra
Kalinga Institute of Industrial Technology, Deemed to be University, Bhubaneswar, Odisha, India

4.1 Introduction

As part of the country's critical infrastructure, healthcare facilities, large and small, must act proactively and quickly to protect themselves from cyberattacks that can directly impact the health and safety of patients and society. According to healthcare professionals with experience in cybersecurity readiness, cyberattacks are identified as the top threat in annual vulnerability analyses for many healthcare systems. In response, the federal government, along with partners in the public and private sectors, continues to work hard to protect the healthcare industry from growing cyber threats [1]. The US Department of Health and Human Services Office of Strategic Preparedness and Response (ASPR) has sponsored ASPR's Center for Technical Resources, Assistance, and Information Exchange (TRACIE) since 2015 [2]. ASPR TRACIE's goal is to fill the gap in healthcare. System readiness capabilities provide a timely and innovative way to share information and best practices during planning efforts. ASPR TRACIE developed this resource to help healthcare providers and their systems to understand stakeholder roles and responsibilities before, during, and after a cyber incident.

Although the focus of this white study is on failures associated with large-scale cyberattacks, many of the strategies and principles outlined apply to a wide range of cybersecurity and medical incidents. To ensure cybersecurity readiness, healthcare facilities and information technology (IT) teams should incorporate key principles of IT readiness into their planning protocols. These include:

- identification of the vulnerabilities being faced by the organization and to develop a plan to address the threats;

Securing Next-Generation Connected Healthcare Systems
DOI: https://doi.org/10.1016/B978-0-443-13951-2.00006-4

- creation of an incident response plan, practice it, and update it regularly;
- incorporating digital cybersecurity infrastructure checklists into operational protocols;
- collaborating with the enterprises and facilities, emergency managers, and IT professionals;
- implementation of a cyber hygiene program to use the cyber hygiene services and employee training sessions to prevent successful attacks;
- identification of clinical and nonclinical operational vulnerabilities within facilities;
- identifying and understanding how to engage with critical external partners such as healthcare coalition (HCC) stakeholders;
- effective mitigation of cyberattacks making it rely on careful planning by the facilities or health system's IT team in conjunction with facility leadership, providers, and ancillary departments; and
- comprehensive routine evaluations of the facility, or health system, across departments and systems can provide insight into their interdependencies and expose vulnerabilities that should be addressed.

A health gadget's first line of protection is the information system (IS) structure that protects the infrastructure and ambitions to lessen the effect on core competencies and capabilities whilst an attack occurs. Within larger health systems, an organization—a huge answer has probably been installed through a group of skilled clinical and nonclinical IT specialists. These answers aim to insulate the device from assault and restrict its spread across multiple systems or programs.

Medium and large healthcare centers should have the right security configuration control protocols in vicinity. The Health Region Council Cybersecurity aids cybersecurity practices for medium and huge fitness care companies and gives records specific to those entities. Separate, probably smaller, related facilities need to ensure that IT cybersecurity tactics are in line with the agency gadget and will follow identical safety protocols and requirements. Those centers should also take certain measures so that they can disconnect from relevant or organizational systems and run independently, to both shield themselves and the principle network should an incident arise. The Health Quarter Council Cybersecurity is useful resource for small healthcare agencies, offering facts specific to assisting smaller centers. Implementing effective cyber hygiene practices is critical to protecting your organization's networks and resources.

Healthcare facilities, particularly smaller ones, with limited IT resources can take advantage of free cybersecurity services and tools provided by the

Figure 4.1 Cybersecurity in healthcare.

public and private sectors, as well as a federal agency (Cybersecurity and Infrastructure Security Agency [CISA]). To help identify system vulnerabilities, assess resilience, and stay abreast of cyber practices, rural health centers can use a toolkit designed specifically for hospitals and clinics in remote areas (Fig. 4.1).

4.2 Quality Function Deployment and background study

QFD technology originated in Japan in the late 60s and early 70s. The Japanese have created a methodology that supports the development of complex products by linking the planning elements of the design and construction process to the specific requirements of the customer. The deployment of quality features expresses the voice of the customer [3]. A set of customer requirements with each requirement prioritized to indicate its importance to the customer. This methodology was applied to Mitsubishi in 1972. In the 80s and 90s, it was gradually and successfully adopted by American and Japanese companies. In recent years, the scope of QFD methodology has expanded and its popularity has grown significantly. There are no limits to the potential applications of QFD.

The essence of the original QFD changed into extracting the prioritized customer needs or dreams, expressed in his/her own phrases (WHATs), to translate them into prioritized technical product first-class traits (HOWs) and ultimately into components' characteristics, running choices, and other decisions. Every translation of client's voices and next methods makes use of a matrix relating the HOWs with the WHATs, associated with any unique QFD degree [4]. The HOWs of one matrix emerge as the WHATs of the following. Vital parameters within the translation manner are the numerical values of the matrix elements representing the energy of the family members between the variables concerned. The QFD method is implemented through sequential matrices that interpret the purchaser requirements expressed in his/her very own words into measurable product technical traits called "The house of great".

The House of Quality is presented in Fig. 4.2. The matrix inputs (the house's western wall) are the patron desires, the WHATs, and their respective numerical importance to the customer. They are translated into the HOWs (the house's ceiling), which represent the measurable product technical characteristics or specs. The relationship between each technical feature and each consumer need is the core of the matrix and displays how properly each technical function expresses the related consumer's wants [5]. The standard relationship strengths are susceptible, sturdy, and very robust and they are all fantastic. The triangle (the house's roof) describes the relationships among the technical characteristics, among which a few are positively associated, even as others are negatively associated and are used as trade-offs. The translated values (the matrix output) represent the calculated significance of the technical traits. As cited earlier, the output of matrix I turns into the input of matrix II [6]. This sequential method is carried on from matrix to matrix.

In this original version of QFD as a product design method, there is one numerical input, the customer needs with their respective numerical

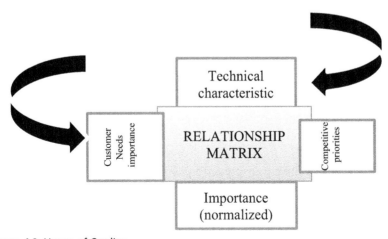

Figure 4.2 House of Quality.

importance. This input is introduced in the first matrix and translated along the entire sequence of matrices.

As mentioned earlier, QFD has evolved into a general all-purpose framework. Barad and Gein apply it to exhausting research aimed at extending established synthetic techniques to help improve manufacturing systems in small manufacturing organizations (SMEs) [7]. They first describe a hard strategic addition for manufacturing firms that includes typical competitive advantages found in the manufacturing strategy literature: time, state-of-the-art, and cost. To do this, they bring all the other benefits to the whole, namely human benefits, which demonstrates the importance of employee participation. In accordance with Flanagan, the desire to develop a human-centered machine stems from a concern to express unfulfilled desires.

This means that stories from the past are more likely to be bad than positive. Therefore, the study does not extract customer needs as a real QFD, but their concerns. To make informed connections between strategic priorities, strategic issues, and work issues, they are all tied to the same overall performance metrics: time, exception, price, and people-centric performance [8]. All questions were rated on a similar severity scale. Here, we apply a technique to represent the study modeling structure and the corresponding QFD matrix. The extended structures of these authors show structural and contextual changes from the original version of QFD [9].

The changes in structure in Fig. 4.3 are as follows:

• The matrices have no roofs.
• Each matrix has two entries. An entry retains the original sequential path. In other words, the output of a given matrix (HOW) becomes

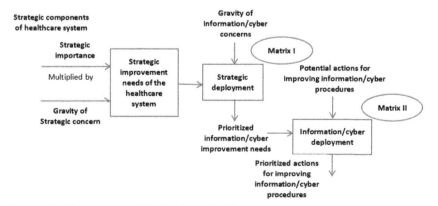

Figure 4.3 The conceptual Quality Function Deployment model.

the input of the next WHAT matrix. The second entry is the COMMENT weight. In the original QFD, HOW weights were calculated by multiplying the WHAT weights by the strength of association with HOW. Since HOE has input weights, the calculation method for the output weights is the same as the original QFD, but the calculation formula also takes into account the input weights of HOE. Another change has to do with the theme and the context. Many topics can be explored in detail besides "product," which is the main subject of product planning, design, and manufacturing.

Considering HOW has input weights, the output weights are calculated in the same way as the output weights of the genuine QFD, but the calculation system also considers the input weights of HOW. In an ongoing study, 94% of fitness organizations said that they had experienced a cyberattack. To date, cybercrime against fitness and healthcare has manifested itself through four unique threats: loss of facts, financial theft, attacks on scientific equipment, and attacks on infrastructure [10]. Some cybercriminals are driven by financial advantage, while others seek to use "hacktivism" to acquire high-profile assets or customer information, damage an organization's profile, or make a political statement. Privacy and security rules introduced by the Health Insurance Portability and Accountability Act (HIPAA) have drawn attention to the importance of hiding nonpublic fitness information and provided a regulatory framework that encourages compliance. Still conformity frequently translates into security.

The fact is that most modern HIPAA defense strategies rely on a standard-based technology approach to segregating sensitive data. However, according to this recent research, many attackers bypass this kind of protection and do not need a secret method to do so. There are significant gaps in knowledge and guidance between technical areas and compliance areas that need to be addressed.

Healthcare organizations in particular face significant monetary risk of statistical data theft due to their respective legal liability positions and the volume and variety of data they acquire and store. Of the 16 industries studied by the Ponemon Institute, a research center focused on privacy, statistical protection, and information protection policy; fitness and healthcare performed best in preserving value files in the case of information loss: approximately $233. Proposals for all industries rose to $136, with retail coming in at $78, according to the document. Costs include prison relocation, rehabilitation, investment in security management, and long-term credit rating protection services for victims. If these anticipated costs apply to the 2009–10 data breach, in which a security breach exposed the personal and health facts of more than 600,000 health plan participants, the total cost could reach billions, forcing WellPoint to pay HIPAA. In healthcare, 72% of recent malicious traffic, viruses, and similar attacks targeted hospitals, clinics, large group practices, and individual providers, with the remaining 28% targeting provider organizations and health plans, health insurance, pharmaceutical, and other companies. In other words, healthcare delivery is proactive and focused [11].

In addition to finding ways to defend against information theft, the healthcare industry must also be alert to other cyber threats such as currency theft, interference with clinical equipment, and attacks on critical infrastructure. Financial robberies are often similar to factual robberies and require similar protections. Ukrainian and Russian hackers, with the help of more than 100 accomplices in the United States, have now stolen $1.03 million from the payroll account of a Washington nursing home, suggesting that a recent spate of such attacks is coming to an end. But threats to clinical equipment and critical infrastructure can even pose an additional challenge as they affect patient health and safety. Patients are especially vulnerable to attacks that can disrupt clinical infrastructure, disrupt communications and services, interfere with clinical equipment, alter or tamper with critical data, or render it unusable.

Although the "Internet of Things" that connects physical devices, including patient video display units that incorporate sensors or actuators and are electronically programmed, enables remote and distributed access to many diagnostic skills and therapies in health facilities, this connectivity also creates opportunities for attacks. In 2009, the Department of Veterans Affairs discovered a massive infection of devices with laptop viruses and other malware. Consequently, not only patient protection was compromised but also the disruption and costs to patients and providers are significant.

A dramatic increase in cyber intrusions and attacks on clinical devices has caught the attention of regulators. In June 2013, the FDA issued a security briefing titled "Cybersecurity for Clinical Devices and Clinic Networks," and security briefing fees for security agencies that included representatives from the FDA, Office of the National Coordinator for Information Age in Health Care, and the Federal Department of Health. Communications Launched a brief calling for better nonpublic sector engagement and an opportunity-based comprehensive regulatory framework—but no longer describes this framework. Unfortunately, heavy and slow regulation can significantly increase costs and potentially mask emerging threats. It would therefore be wise for the medical community to hear the call from regulators and actively engage in the dialogue. Ponemon research shows that organizations that adequately focus on improving their cybersecurity posture, hiring and empowering an information security officer, and building robust incident response systems can reduce the potential financial risk of a data breach by 42%. An organization's security system includes a combination of skills, operations, and processes that are often really hard to understand, let alone develop. At some point, it is not always clear whether the focus should be on modernizing technology, employee training, physical security, or a combination of these. None of these tasks are easy, cheap, or quick; prioritization and a clear strategy are essential [12].

An active learning approach is required to focus and succeed with prioritized cyber protection strategies and tactics. This harvester must be able to understand the complex interplay and dynamics between external threats, inherent vulnerabilities, specific risks, and system resilience, all of which must be understood within the specific context of the healthcare delivery environment. Although we cannot predict exactly what our opponents will do, we can control our own environment and we must keep an eye out for potential opponents. Just as public health strategies have been developed to detect and track emerging epidemics, identify population risks and vulnerabilities, and prevent or mitigate adverse effects, similar approaches can be used to improve cybersecurity in healthcare delivery organizations (Fig. 4.4).

Figure 4.4 Representation of healthcare business.

First, proactive, real-time monitoring and communication of emerging cyber threats can be used to analyze threats and ultimately influence public policy and prevention. Second, risk-based analysis and modeling that considers current and potential threats, resulting risks, and the vulnerability and resilience of ISs can guide policymaking. Third, effective regulation helps ensure the fidelity of medical devices; striking the right balance in establishing safety without creating another set of costly and distracting compliance standards will require stakeholders (patients, providers, and institutions) to predefine—perhaps in a forum organized by the Institute of Medicine, based on its data privacy and security reports. The threat of cyberattacks is evident in healthcare. Now is the time to organize, come together, and focus in ways that truly protect our patients, our providers, and our institutions. Technology has undoubtedly improved healthcare. Let us make sure its promised benefits are always delivered safely. Ahmed et al. conducted a review of security metrics and proposed a reference architecture for aggregating the security of an enterprise network [13]. Jafari et al. discussed security metrics in eHealth and proposed an approach that consists of five elements: technology maturity analysis, threat analysis and modeling, requirements establishment, policies and mechanisms, and system behavior, but the method for developing the metrics was not described [14]. Liu et al. focused on the implementation of IPsec to protect the confidentiality and integrity of sensitive patient data from cyber threats and demonstrated how to implement the controls [15]. Abie and Balasingham proposed a risk-based adaptive security framework for IOT in eHealth that allows systems to learn and adapt to changes in the environments by anticipating threats [16]. Savola et al. performed risk analysis on an eHealth self-care system and quantified the metrics using a risk assessment technique based on severity and impact [17]. Muthukrishnan et al. proposed a quantitative and qualitative metrics maturity method based on a scorecard [18]. Tounsi et al. categorized indicators of compromise (IOC) into network-based indicators (IP addresses, URLs, and Domain names) [19]. Catalogue et al. proposed an automated technique for collecting IOC from web pages in a honeypot setup [5].

4.3 Security measures

- Security metrics have been gaining interest from researchers in academia and the industry to help quantify security measurements and assist with decision-making. According to a recent report by Thycotic, most organizations are failing to implement cybersecurity metrics and, therefore, are unable to evaluate and track the effectiveness and performance of their security mechanisms. Several authors have published work on security metrics but these were mainly targeted at an organizational level and not captured in the context of healthcare IT systems.

• In the following subsections, we present some security metrics that could be used to address the cybersecurity challenges discussed in the previous section and describe how these metrics could be used to improve the security of IT systems. We believe that the metric categories we are proposing have a wider coverage and could be used to measure and improve the security status of the IT systems including interconnected devices. Although these metrics could be applied to any organization, we believe their positive impact will be felt more in the healthcare sector where the attack surface is much larger due to the vast number of legacy systems, interconnected medical devices, complexity of the endpoints, security culture, and the high value of the data they hold.

We categorized the security metrics into eight groups, in Fig. 4.5, which are: IOC, indicators of attacks (IOAs), resilience, red and blue teaming, vulnerability assessment, intelligence-driven defense, risk assessment, and penetration testing. In the following subsections, we will describe each of the categories in more detail.

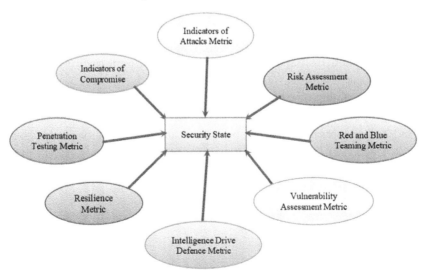

Figure 4.5 Security metrics.

4.3.1 Indicators of compromise-based metrics

IOC are information that can be used to identify malicious activities that have taken place on a system or network. The information relating to these security breaches can be used to prevent similar attacks in the future. Monitoring for IOC can help organizations to quickly detect incidents that may have been missed by the monitoring tools. Incidents such as malware attacks often take a long time before they are detected by the affected organizations or by the security community and by that time lots of sensitive data may have been stolen. Once the IOCs are detected and shared, organizations could act on these indicators to mitigate the risks and prevent future attacks.

Sharing cybersecurity intelligence such as IOC could allow organizations to defend against sophisticated cyberattacks. For example, in the case of WannaCry, there were three IOCs, which were detected following the first wave of attacks. One of these was a dropper that contained the ransomware and was used to run it while the other two were encryption plugins.

Data relating to IOC such as artifacts that are left after malware executions could be obtained from the log's files. Monitoring tools such as antivirus systems are known to use IOC to block malware. Tounsi et al. categorized IOC into network-based indicators (IP addresses, URLs, and Domain names), host-based indicators (malware names, signature, and registry keys), and email-based indicators (source IP, header, attacked link).

There are free tools, such as redline from FireEye, that can be used to perform IOC analysis on systems. Other web-based tools such as IOC Bucket and IOC Editor can be used to share, research, create, and edit IOCs. Catalogue et al. proposed an automated technique for collecting IOC from web pages in a honeypot setup.

Metrics based on IOC could assist IT professionals including those in healthcare to learn more about the security threats affecting their systems and to use the lessons learnt to prevent future attacks. In the next points, we will mention some useful metrics relating to IOC and point out how they could be used by security professionals to enhance the protection of their systems. These metrics are given in Fig. 4.6.

- *Volume of outbound traffic.*

    An unusual increase in outgoing traffic could be a sign of data exfiltration. Organizations have a responsibility to protect their sensitive data and any breach is likely to attract attention from the government bodies with responsibilities for enforcing the data protection laws

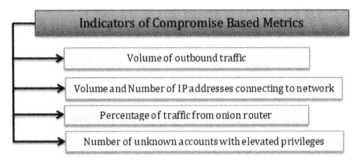

Figure 4.6 Parts of indicators of compromise-based metrics.

such as GDPR. The large volume of personal data held by healthcare providers is one of the reasons why cybercriminals target this sector. Using this indicator of compromise will allow healthcare IT professionals to look for incidents that may have been missed by the security monitoring tools. It is paramount that IT professionals correlate log data from affected systems and the other network monitoring tools to determine exactly what happened including the type of malware.

• *Volume and number of IP addresses connecting to your network from outside of your geographical area.*

Successful logins from outside your region could be an indication of a security breach. Given the complexity of the endpoints and the legacy systems that are still in use in healthcare institutions especially in public hospitals, there are more chances of breaches not being detected compared to other sectors such as the financial sector. Hunting for IOC such as external IP addresses from outside of your normal geographical area could allow the IT security professionals to find out whether successful login was established and to investigate the incidents further. Number of simultaneous logins by the same user from different locations that has not been detected by the network security tools could be an indication of sophisticated attacks that bypass the security controls. Investigating this IOC will help IT security professionals determine the chain of events that led to this breach and fine-tune their security controls to help detect or prevent similar attacks in the future.

• *Percentage of traffic from onion router that was not detected by the network monitoring tools.*

You can use these metrics to test the effectiveness of your security controls. Cybercriminals are known to hide their identities by moving data between multiple Tor nodes, making it difficult to track their activity. IT security teams should monitor these metrics and put in place appropriate security measures to mitigate the risks associated with them.

- *Number of unknown accounts with elevated privileges found on compromised systems.*

This can be a sign that the system has been compromised and that an attacker has performed a privilege escalation to take ownership and create a backdoor.

4.3.2 Indicators of attack metrics

IOAs are proactive security measures that can be used to detect attacks before IOC are displayed. IOAs allow organizations to prevent attackers from taking advantage of their systems by implementing mitigation controls. For example, IOAs can be used to prevent attacks, such as phishing and ransomware, that have become popular attack vectors for healthcare providers. Implementing IOA indicators such as those shown below can help organizations bolster their proactive approach and prevent attacks that could compromise their systems. These metrics include the parts given in Fig. 4.7.

- *Lateral movements.*

This could indicate an attacker has gained access to the network and is moving from one vulnerable host to another until the goal is achieved. Due to the complexity of the setup and the number of interconnected devices in healthcare institutions, IT security professionals will need to monitor this metric to enable them to act promptly and implement corrective measures to disrupt such activity.

- *High Bandwidth and increased outgoing traffic.*

This could be a sign of a DDoS attack or data exfiltration.

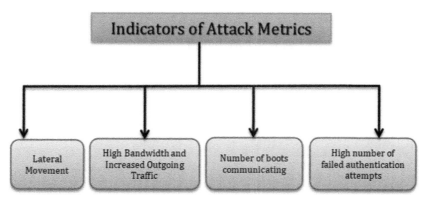

Figure 4.7 Parts of indicators of attack metrics.

- *Number of hosts communicating with external networks on nonstandard ports.*

  This could be an indication of sophisticated attackers hiding their activities. An attacker could use nonstandard port numbers to avoid detection by the security controls. Monitoring this metric could help security professionals to stop unauthorized connections and to update the secure measures.
- *High number of failed authentication attempts.*

  This could be an indication of an attacker attempting to gain access to your network using brute force. Number of processes with high memory consumption could be an indication that malware is being executed. Malware usually runs on the memory, and security professionals should investigate spikes in memory usage to determine whether malware execution is taking place.

4.3.3 Risk assessment metrics

Understanding what needs to be protected is the first step in the risk assessment process. Asset identification and classification will need to be performed in order to direct resources to the most critical assets. Healthcare institutions hold lots of sensitive data relating to their patients and it is important for all assets to be accounted for and their impact on the business quantified. A risk register should be maintained and populated with risks that were identified along with any corrective measures taken and should be regularly updated. Organizations can use security metrics to maintain oversight of their risk management processes. These metrics in Fig. 4.8 include:

- *Percentage of risks with severe or critical rating.*

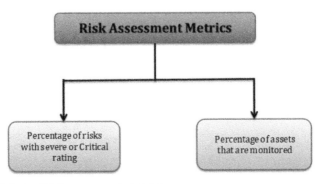

Figure 4.8 Parts of risk assessment metrics.

This will help to prioritize and direct resources to deal with the most urgent threats affecting critical assets. The healthcare sector, especially public hospitals, has limited resources, and monitoring these metrics will enable them to prioritize and deal with risks with the highest severity. Percentage of key assets with no ownership assigned: Asset ownership is key to determining the right safeguards needed to protect these systems. Given the ever-changing threat landscape and the presence of legacy systems in the healthcare sector, it is just a matter of time before such systems are compromised. Resuming services and invoking recovery plans for such systems depend on the asset owner if the system is to be restored with limited downtime. Organizations should ensure all assets have an owner.

- *Percentage of assets that are monitored.*

This metric will help with monitoring the coverage and determine if gaps exist. Considering how critical some of these systems are to service delivery and their complexity in healthcare settings, efforts should be made to ensure coverage is maximized.

4.3.4 Penetration testing metric

Penetration testing is a method for detecting security vulnerabilities on a network and can involve simulated attacks. Due to the costs involved, organizations have the tendency to perform penetration testing on a quarterly or biannually basis but the threat landscape could change very quickly and make such tests redundant within a short period. To fill this gap, newer penetration testing tools that operate inside the network have been developed by companies. For example, fire drill developed the attack IQ, which is a tool that performs automated testing to determine the enterprise security posture, but these tools are not yet widely adopted. Healthcare institutions have a large attack surface due to the vast number of endpoints and legacy systems. Performing regular penetration testing on these systems will enable the organizations to detect weaknesses and implement the necessary controls to prevent potential attacks. One way to maximize these tests is to use penetration testing metrics. These metrics in Fig. 4.9 include but are not limited to:

- *Percentage of penetration tests that discovered high risks.*

  This can indicate a measure of how well the existing security controls are performing including the detection capability of the vulnerability management tools.

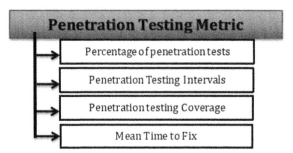

Figure 4.9 Parts of penetration testing metric.

- *Penetration testing intervals.*

  The threat landscape changes over time and increasing the frequency of testing will help organizations to uncover security risks and help prevent potential attacks. It can also be used to measure how well vulnerabilities detected in the previous tests were resolved.
- *Penetration testing coverage.*

  In healthcare settings, the coverage metric should include all critical systems such as medical devices and the infrastructure. Ideally, all assets should be covered.
- *Mean time to fix (MTTF).*

  These metrics will show the average time taken to fix the vulnerabilities that were identified during the penetration testing and also allow senior managers to measure the capability and average response times of the technical teams.

4.3.5 Vulnerability assessment metric

Vulnerability is a weakness on a system that could be exploited by a threat. where possible organizations should be implementing automated vulnerability scanning tools to enable them to detect vulnerabilities more efficiently and in a consistent manner [20]. System administrators should monitor and deal with the detected vulnerabilities promptly according to their severity and impact. The vulnerability assessment tools are likely to detect a large number of vulnerabilities in healthcare institutions such as public hospitals. One way to maximize the effectiveness of the response is to use vulnerability metrics and such metrics in Fig. 4.10 include:

- *Percentage of critical systems with known vulnerabilities that are not patched.*

  This metric will help organizations to determine how well they are implementing their security patches. Systems with known vulnerabilities

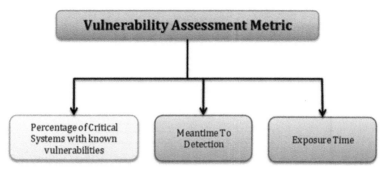

Figure 4.10 Parts of vulnerability assessment metric.

face a higher risk given this information is publicly available on the vulnerability databases. The small and medium-sized business vulnerability which was exploited by the WannaCry ransomware is one example of known vulnerabilities which could have been prevented with a simple patch.

- *Mean time to detection*

    This metric will enable organizations to test how well vulnerabilities are detected including the effectiveness of their vulnerability management systems and other security controls.

- *Exposure time*

    This is the time between the discovery and application of patches and can be used to measure how quickly administrators apply patches. The sooner you apply for a patch, the better.

4.3.6 Red and blue teaming metric

Red teaming is a simulated form of attack in which skilled teams attempt to penetrate the security defenses and compromise the systems. Organizations usually employ the service of red teams in order to test the maturity of their security controls. After the end of the red team assessment, organizations will have a list of attack vectors they are vulnerable to and corrective measures to mitigate such risks [21].

Red teams should be complemented with blue teams whose role is to defend against the attacks and bolster the security defenses. The purpose of a blue team is to defend the organization against both red teams and real attackers. Enlisting the service of red teams can be expensive and organizations should try to maximize this potential to improve their security.

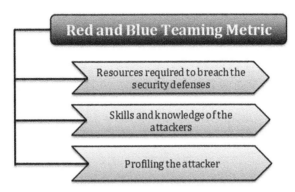

Figure 4.11 Parts of red and blue teaming metric.

Red and Blue teams could also be used during the op–iteration stages after new security programs are deployed. organizations can use metrics to measure how well they are integrating the outcome from the red team assessment to improve the strength of their security mechanisms. These metrics are given in Fig. 4.11 include:

- *Resources required to breach the security defenses and compromise the systems.*
  This will indicate how well the systems are protected and the layers of defense. The security defenses should be strong enough to withstand most attacks and the more resources required to break in the better the defense mechanisms.
- *Skills and knowledge of the attackers.*
  This metric can be used to determine the expertise and sophistication required to break into the systems and the information obtained from this metric could be used to increase the security strength.
- *Profiling the attacker.*
  System administrators could use honey-nets to build an overall picture of the attacker. This metric could be used to put in place mitigation controls and to beef the defensive capabilities.

4.3.7 Resilience metric

Resilience is the ability for a system to adapt and continue to provide functionality in face of an attack. The healthcare sector experienced a vast number of targeted attacks which disrupted critical services as seen with the WannaCry attack on the UK National Health Service (NHS) [22]. Ransomware attacks such as WannaCry encrypt the files and make them unavailable to the users until a ransom is paid. Resilience will enable

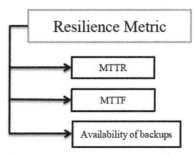

Figure 4.12 Parts of resilience metrics.

organizations to withstand adversarial attacks and to ensure continuity of critical services. The following are some of the metrics that could be used to measure resilience in an organization given in Fig. 4.12.

- *Mean time to repair*

    This metric will enable systems administrators to monitor how quickly they respond to incidents that disrupt availability of their critical resources and restore the service to normality. In healthcare, unavailability of critical resources can have a serious impact on patient health. During the WannaCry ransomware attack, the affected NHS hospital had to cancel lifesaving operations and, in some cases, divert ambulances too far away to hospitals that were not affected by the ransomware.

- *Mean time to failure*

    This metric could be used to measure the resilience of the systems in terms of the frequency and length of time between failures. Reliability of systems is critical to most organizations, but it is more so in the healthcare sector given the number of legacy systems in use and the impact it will have on service delivery.

- *Availability of offline and tested backups*

    This metric could be used to measure the reliability of the backups. Sometimes backup might be the only option to restore services in some cases such as ransomware attacks. Organizations with a tested offline backup are less likely to end up with encrypted backup, given malware developers are known to target online back to force victims to meet their demands.

4.3.8 Intelligence-driven defense metric

Cyber threat intelligence is gaining popularity for its proactive approach. Security professionals use threat intelligence to learn more about intruders targeting their area. Threat intelligence often depends on the existence of a vulnerability and the availability of threats capable of exploiting it.

Threat intelligence models widely used in the industry include Cyber kill Chain, Diamond Model, OWASP, and Attack Graphs. The cyber kill chain is an intelligence-based defense approach that can be used to prevent an attack by disrupting an attacker's activities at one of the seven stages described in the model. The Diamond Model enables security professionals to understand the behavior and capabilities of intruders and has four characteristics, namely: infrastructure, capabilities, adversary, and victim. A cyber kill chain can be used to quickly determine the severity of an attack by mapping events generated by various sensors. For example, an event from a host intrusion prevention system might mean that an attacker has moved past the early stages of the attack process, in the installation or operation phase, and must therefore be assigned a high priority to handle such an attack [23]. Examples of threat intelligence metrics that could be used to enhance protection are:

- *Threat intelligence teams*

  This metric can be used to measure the capabilities of technical resources. Your in-house threat intelligence team can review internal and external threat intelligence databases and weed out false positives.

- *Number of known threat groups targeting your organization or sector at any given time*

  This metric can be used to gauge how well these communities are able to combat threats. Historical data can be used to analyze behavioral patterns. Data related to these groups can come from internal intelligence groups or vendors and the security community as a whole.

- *Number of attacks detected and mitigated using the attack models such as the cyber kill chain*

  This can indicate how well your threat intelligence team is doing. Information from these metrics can also be used to protect your organization from future attacks as part of the lessons learned.

- *Access to vendor threat intelligence reports directly related to your organization*

  Vendors have vast capabilities and resources, including threat sharing with industry partners. Access to intelligence feeds of these threats can give organizations an edge over attackers.

- *Metric from IOC*

  This metric was discussed in the subsection above.

- *Knowledge of online forums where exploits are sold or discussed*

This metric is no longer just an indication of how active your security team is, but it is also one of the best ways to determine how risk actors are using more modern strategies to exploit vulnerabilities (Fig. 4.13).

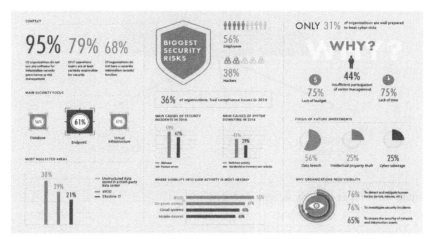

Figure 4.13 Information technology risks.

4.4 Cybersecurity wellness

Internally, organizations should conduct regular drills to ensure that stakeholders, vendors, and emergency management are prepared for a cyber emergency. Involving medical personnel and ancillary components in these scenarios is integral to understanding closure procedures in both clinical and nonclinical settings. Although exercise frequency and intensity are not uniform across healthcare organizations, the goal of exercise is to identify gaps in current cybersecurity practices and identify areas for improvement. To optimize the results of the exercise, the plan should include establishing a timeline for revised results to improve current protocols. The Homeland Security Exercise and Evaluation Program provides guidance for the development, execution, and evaluation of exercises that address readiness. CISA provides a resource list for the Desktop Exercise Package (CTEPS), which includes a library of cybersecurity scenarios [24]. The various exercises are shown in Fig. 4.14.

4.4.1 Use case scenarios

* *Develop exercises with varying degrees of impact levels*

   Developing exercises that mimic real-world cyber incidents and address individual unit/office responses. Many exercises are restricted to specific systems, applications, or emergency situations, however, postattack summaries often reveal that limited exercises often do not align with the implications of a real cyberattack.

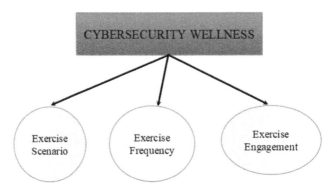

Figure 4.14 Types of cybersecurity wellness.

- *Run exercises for an array of scenarios*

    Focus on impacts to mission-critical applications and subsequent effects on healthcare operations. Practice with 1−2 compromised systems and move toward larger-scale cyber incidents that impact the entire organization (worse-case scenarios). Consider utilizing a white hat (hired cybersecurity expert) to attack and stress test systems.

- *Explore use of third-party IT specialists*

    A purple team, to facilitate exercises aimed at identifying vulnerabilities and providing solutions as part of an independent assessment.

- *Conduct specific drills*

    Conducting specific drills for paper charting and manual clinical processes, especially for novice, new, and younger staff who may not be familiar with older manual processes. To alleviate the steep learning curve, practice hands-on and in-real-time activities versus solely providing handouts.

- *Incorporate communication challenges into scenarios*

    Run drills for internal communication issues (e.g., no phones, email, and paging systems); practice answering staff and public-facing questions (e.g., news media, patients, and social media).

4.4.1.1 Exercise frequency

- At the health system level, routine exercises should occur one to two times per year with individual units testing their specific workflow at least twice a year.
- At the department level, continuously reinforce cyber hygiene practices during team meetings, educational series, and other appropriate forums.

- Once per year, organizations should run full tabletop exercises for the entire health system.

4.4.1.2 Exercise engagement

- Consider establishing a subcommittee or specialized group within the hospital to focus on IT security and downtime response needs to represent disparate interests within the health system.
- Healthcare providers, IT, emergency management and biomedical representatives, vendors, and law enforcement must participate in a training program. Provide summaries of situational awareness and lessons learned, as appropriate.
- Ensure nonmedical departments and staff, such as public relations and communications representatives, are included in drills.
- Identify employees who may be affected by restricted access to critical technology (e.g., computer, internet and software application) after a cyberattack. Include them in drills to determine equipment and supply needs to ensure business continuity and identify and resolve workflow issues.

4.5 Response strategies

When a cyber incident is suspected, IT specialists will immediately begin determining the extent of the impact on each of the systems or infrastructure. While they investigate the extent of the damage and take action to isolate, repair, or remove the 4,444 affected pieces of technology, the hospital will activate information management technology (IMT) to stabilize operations and maintain safe patient care [25]. The following steps can help ensure a smooth initial response:

- Prompt reporting of incidents to appropriate authorities (e.g., CISA and FBI) will allow for prompt delivery of Federal, State, and local assistance/response services to affected facilities.
- The IMT (or similar peer) will scale the event to initiate and handle the correct response. These efforts should be based on predefined severity/impact definitions.
- The level of system downtime will correlate to the scope of impact on the IT infrastructure.
- Every healthcare system and facility should be aware of the thresholds that lead to system shutdowns. Delegating IT teams and authorizing

the shutdown of one or more systems as soon as the threshold is reached is critical to mitigating further damage.

- Implementation of business continuity plans/COOP upon activation of an IMT.
- Including ancillary representatives and partners in all aspects of the response effort.
- Following all the established protocols for integration and periodic communication with HCCs, healthcare system command centers, and other emergency management groups as appropriate.
- Ensuring compliance with enterprise, federal, or SLTT disaster response requirements.

4.5.1 Response downtime procedures

Departments must be prepared to quickly transition to using manual downtime forms and mapping processes during cyber incidents. Healthcare system IMTs must continually estimate downtime and adjust response procedures as needed.

4.5.2 Downtime forms

- Document any new or updated solutions immediately and distribute them to management/staff so that the protocol can be updated throughout the organization.
- In cases where it is difficult to plan an outage due to the severity of the cyber incident, prefer to establish a longer outage procedure to reduce ambiguity.
- Establish a process-savvy closing team to support staff with the time-consuming distribution and implementation of closing documents.
- Confirm that the shutdown documents used are up to date; if new forms need to be generated, ensure that they go through the appropriate approval channels.
- Ensure that, when completing the form, the handwriting is legible and the information provided is correct and includes all required data points.
- Consider security requirements for records, documents, charts, and forms that may contain personally identifiable information or be subject to HIPAA compliance. Protects all paper data and financial information.
- Ensure new shutdown procedures meet regulatory requirements for data submissions, website access, etc.

4.6 Safety considerations

- Ensure good communication between IT response teams and departments to report security issues as they arise. Create a robust incident reporting mechanism with tracking capabilities. Ensure employee is aware of reporting protocol.
- Safety officers, patient advocates, and case managers should visit patient care areas for routine monitoring to proactively identify areas for improvement and provide safety reminders.
- Report adverse events with patient impact to appropriate managers if necessary. Create a workflow to share, distribute, and collect security forms for incident reporting (e.g., Microsoft Form) and identify a repository to process and store these reports.
- Avoid pharmacy ordering incidents by ensuring downtime forms include the required safe components (dose range, appropriate unit, frequency, schedule, and dosing route). During an outage, most security incidents are related to medical orders that do not contain all the necessary information. Please note that drug interactions and allergy alert software/validation may not work and increase potential risk.
- Make sure new/less experienced clinical staff know which components are required for ordering. Without the EHR referenced as a decision support tool, highlighting key information and standardizing order sets, important information may be overlooked. Consider using pre-printed command sets (e.g., Sepsis, Diabetes Ketoacidosis) for high volume/high consequence commands [26].
- Create tables for high-risk medications/all medications with associated protocols (e.g., insulin drops).
- Demographic information may not be included in outage reports (e.g., BCA report) substitution skills were identified to validate the patients.
- Contact a third party by telephone to facilitate discharge to a qualified nursing facility for approval and plan accordingly.

4.7 Facility security considerations and recovery

During a cyberattack, normal security services may be altered or unavailable. Changes to the security protocol should be addressed immediately.

1. Know the location of all controlled access points, provide keys to necessary personnel, and recruit additional security personnel to monitor locked units. Involve hospital security and/or local law enforcement.

2. Considering the following measures to meet the need for additional officers/staff (Stationary or Rounded) in special security/restricted access areas:

- Mother and baby tracking solutions in maternity wards, deliveries and nurseries are planned.
- Establish alternative connection and security protocols in psychiatric hospitals.
- Additional security personnel at gates and in entry/exit areas.
- Determine if visitors should be restricted.
- Develop a plan to secure and provide access to medicine cabinets.
- Ensure time sheets are used as instructed and recorded at the end of the shift.

The severity of the cyberattack will inform the duration of the recovery time. Although recovery is immediate at times, long-term recovery and return to normal operation will involve continuous analysis and adjustments to the system will take longer. Every time a change is made, the department must monitor the safety and security operation. Recovery efforts across all sectors can be resource-intensive. If possible, verify that the attack no longer occurs or does not continue to occur.

- Think of recovery as an ongoing part of downtime. As departments and systems recover, assess the level of functionality that would be beneficial or detrimental to operations (for example, if the system is only partially functional, will the missing functionality impede the flow of work or increase the risk?)
- After planning system recovery, manual documentation is restored in electronic format. Carefully consider the downstream impact on employees during this difficult process (even if they still have to do their "day jobs").
- Available staff will continue to provide additional support to assist with in-depth data entry and card reconciliation.
- IT staff and others may continue to work long hours to discover, test, and fix persistent system issues now that they are back online.
- Can specify that some manually recorded data should not be reconciled with the EHR; must save and share the description of the type of information and how it is stored.
- Some computer applications may be considered unrecoverable. In this case, a new replacement capacity must be deployed.
- Labor management needs to plan for staff mobility and how it will meet labor needs over an extended period of time. Consider the long-term

impact on the workforce (e.g., parenting, mental health, and spousal/partner support). Determine where to find additional support staff.

- Determine if vendors can help restructure biomedical records and meet additional salvage needs for specialized biomedical equipment.
- Resume any suspended diagnostic or therapeutic procedures as soon as conditions permit. Establish a process to contact patients with delayed outpatient appointments and make arrangements to reschedule plans to repatriate transferred patients who wish to return in due course.
- ITM should work with communication leads and establish a process to provide updates to external stakeholders. Determine the report format, which is part of the "need to know" group, and how often updates occur. Use simplified language and avoid technical or clinical jargon. Make sure the message meets the information-sharing restrictions.
- Retraining may involve reloading/correcting software on fixed medical equipment. Involve clinical engineering staff and service providers in organizing such events.

4.7.1 Financial recovery

Financial turnaround spans the entire revenue cycle. Health information management (HIM) is an integral part of the recovery effort to ensure the integrity of the cycle, from patient check-in to claims processing and collection. In the event of a cyber incident, the finance team will need to determine which recovery activities will be performed by HIM resources. The following steps will ensure that the financial recovery process runs as smoothly as possible:

- Start collecting response-related financial data early and continue to follow the submission process and format as outlined.
- Explore insurance options to help with loss and interruption of income.
- Keep service dates for all accounts within the event time frame.
- Include financial SMEs in turnaround meetings to provide insight into timelines, manage expectations, and plan ahead for ongoing recovery efforts.
- HIM's partnership with compliance officers and vendors is important when coordinating outage tickets/documentation.
  - Begin coding from paper records while waiting for the system to recover in an effort to reconcile gaps in records.
  - Plan to increase staff to do the coding. Consider hiring a vendor to help.

- Implement a documented quality assurance process and report all defects that require immediate review.
- Do not aggressively pursue noncritical defects with vendors to avoid overload, and allow for a reasonable signature wait time.
- Establish a financial policy to complete file closures when signatures cannot be verified. Use a clearance stamp or mark to identify reviewed, reconciled, and closed records.
- Check that dates are marked correctly for completeness of income. Ensure that all manually entered incurred costs are entered consistently into the process, and document the workflow.
- Gather the required claim documents for submission in accordance with the outline prescribed by the insurance company, Foreign Exchange Management Act, or other compensation provider.
- Initiate claims for reimbursement and insurance, if applicable.

4.7.2 Demobilization

- Define the criteria for declaring the incident over and resuming normal operations. Notify appropriate stakeholders, prepare a final press release, and update website, intranet, and telephone services.
- Collect after-action reports, lessons learned, and documentation for corrective actions/improvement of plans. While the incident is still recent, a hot flash during a shift change or at another announced time completed cycles, as determined by the incident commander.
- Identify the repository where postevent data is stored. Include documentation of any new downtime forms or workflows and safety or security incidents.
- Distribute a schedule of postincident activities with deadlines to ensure compliance.
- Provide incident orders and complete the replenishment process.

4.8 Development of specified cybersecurity within healthcare organizations

There are differences in the relative growth of organizations across the six cybersecurity areas measured. In particular, the lowest level of growth was found in knowledge and education. Current research shows that these names are important to cybersecurity. A study of UK schoolchildren's attitudes to online safety found that individuals display the most positive attitudes when they are taught appropriate online safety skills and subsequently confident to be responsible. One explanation for the lack of growth in information and education may be the perception that cybersecurity is a concern of the information and communications technology (ICT) sector rather than frontline workers.

However, cybersecurity as a patient safety concern should be considered for all workers in all healthcare settings. According to the employee's job, right of entry, and cybersecurity profile in the organization, the main points for the education, training, and knowledge of the employee should be determined and established. For example, the cybersecurity knowledge and skills of the staff in the ICT department are different and more than the nurses in the emergency room. But frontline workers can still play an equal role in supporting an organization's cybersecurity. Therefore, the importance of education, training, and knowledge must be recognized. This can be used in many ways but must have clear information and easy access to the study of all.

Survey respondents rated high (3.2) on the growth of cybersecurity technology. Globally, this is somewhat unexpected, as many advances in technology can be expected, especially when comparing medical facilities from high-, middle-, and low-income groups. This finding may be influenced by the nature of the clinics involved, including many private clinics in the sample. These organizations may not represent traditional public hospitals and health technologies found in national health systems. As there is currently no published data comparing the cybersecurity of international health organizations, a few organizations such as the World Health Organization (WHO) should consider creating a cybersecurity document in future eHealth and mHealth studies, such as the WHO Global eHealth Survey published in 2016 [27].

Further work is needed to determine the level of adoption of global health technology reform and how best to secure it, including decision-making regarding medical device manufacturing and related safety standards, to develop a cybersecurity strategy used in the technologies used.

The development work shows that governance and regulation are two areas where players have strong cybersecurity mechanisms. More importantly, the management and control of the six elements of cybersecurity are usually managed outside the hospital or at the top level, taking cybersecurity measures. An important consideration going forward is the regulatory and healthcare environment in which healthcare organizations operate, and how the organization's appropriate cybersecurity governance mechanisms can guide healthcare organizations in guiding their cybersecurity business expansion plans, including creating a culture of cybersecurity [28].

Of course, this is not an easy task. Instead, it requires coordinated coordination and collaboration between health systems, programs, and projects. The first step will be to identify the responsibilities of stakeholders, including its assets (financial and human resources), processes and business processes, users and patients, and the organization itself, including the broader health and cybersecurity space. This expansion will include consulting and advertising agencies, medical device manufacturers, and other external stakeholders associated with the healthcare cybersecurity organization. Once a relevant map is available, an assessment can be made of current and future network arrangements and stakeholders' compliance with these regulations.

Discussion of language should be as important as the open and transparent sharing of medical institutions' best practices in this process. This collaboration between organizations will lead to best practices and expertise in establishing cybersecurity policies in healthcare.

4.9 Conclusion

This article has successfully discussed the security challenges facing the healthcare sector and has given some reasons why the sector is vulnerable to cyberattacks. Cyberattacks are a problem that affects every industry, and they are not just limited to healthcare, with far greater implications when it comes to patient safety. Vulnerabilities in medical devices, infrastructure, and safety culture were discussed. Finally, cybersecurity metrics for enhancing the protection of these systems were proposed. The metrics were grouped into IOC, IOA, penetration testing, red and blue teaming, risk assessment, resilience, and intelligence-driven defense [29]. Future work will involve quantifying and aggregating these metrics to provide a higher-level view of the security status of the networked systems including medical devices and other connected systems.

References

[1] S. Ghafur, G. Fontana, G. Martin, et al. Improving Cyber Security in the NHS [online]. Institute of Global Health Innovation, Imperial College London, 2019. Available from: https://www.imperial.ac.uk/media/imperial-college/institute-of-global-healthinnovation/Cyber-report-2020.pdf.

[2] Joint Learning Network. Using Health Data to Improve Universal Health Coverage: Three Case Studies. Joint Learning Network for Universal Health Coverage, PATH, Wipro Ltd., 2018.

[3] S. Ghafur, E. Schneider, Why are health care organizations slow to adopt patient-facing digital technologies [online]? Health Affairs Blog (2019). Available from: https://doi.org/10.1377/hblog20190301.476734.

[4] G. Martin, P. Martin, C. Hankin, et al., Cybersecurity and healthcare: how safe are we? BMJ (Clinical Research ed.) 358 (2017) j3179.

[5] M.S. Jalali, J.P. Kaiser, Cybersecurity in hospitals: A systematic, organizational perspective, Journal of Medical Internet Research 20 (5) (2018) e10059.

[6] World Health Organization (WHO). WHO reports fivefold increase in cyber-attacks, urges vigilance [Press Release]. World Health Organization; 2020. Available from: https://www.who.int/news-room/detail/23-04-2020-who-reports-fivefold-increase-in-cyberattacks-urges-vigilance

[7] Global Cyber Security Capacity Centre. Cybersecurity Capacity Maturity Model for Nations (CMM) Revised Edition (Online). University of Oxford, 2016. Available from: https://cybilportal.org/wp-content/uploads/2020/05/CMM-revised-edition_09022017_1.pdf. This preprint research paper has not been peer reviewed. Electronic copy available at: https://ssrn.com/abstract = 3688885 P.

[8] W.S. Humphrey, Characterizing the software process: a maturity framework, IEEE Software 5 (2) (1988) 73−79.

[9] R.A. Rothrock, J. Kaplan, F. Van Der Oord, The board's role in managing cybersecurity risks, MIT Sloan Management Review 59 (2) (2018) 12−15.

[10] S. Ghafur, S. Kristensen, K. Honeyford, et al., A retrospective impact analysis of the WannaCry cyberattack on the NHS, npj Digital Medicine 2 (98) (2019). Available from: https://doi.org/10.1038/s41746-019-0161-6.

[11] V.F. Nieva, J. Sorra, Safety culture assessment: a tool for improving patient safety in healthcare organizations, BMJ Quality & Safety 12 (2003) ii17−ii23.

[12] E. Kweon, H. Lee, S. Chai, et al., The utility of information security training and education on cybersecurity incidents: an empirical evidence, Inf Systems Frontiers 23 (2021) 361−373.

[13] S. Pfleeger, D. Caputo, Leveraging behavioral science to mitigate cyber security risk, Computers & Security 31 (4) (2012) 597−611.

[14] World Health Organization, Global Diffusion of eHealth: Making Universal Health Coverage Achievable. Report of the Third Global Survey on eHealth, World Health Organization, 2016, Available from: https://www.who.int/goe/publications/global_diffusion/en/.

[15] C. Nazli, S. Madnick, J. Ferwerda, Institutions for cyber security: international responses and global imperatives, Information Technology for Development 20 (2) (2013) 96−121.

[16] S. Ghafur, E. Grass, N.R. Jennings, A. Darzi, The challenges of cybersecurity in health care: the UK National Health Service as a case study, The Lancet Digital Health 1 (1) (2019) E10−E12.

[17] L. Coventry, D. Branley, Cybersecurity in healthcare: a narrative view of trends, threats and ways forward, Maturitas 113 (2018) 48−52.

[18] C.S. Kruse, B. Frederick, T. Jacobson, D.K. Monticone, Cybersecurity in healthcare: a systematic review of modern threats and trends, Technology and Health Care 25 (2017) 1−10.

[19] M. Hagland, With the ransom crisis, the landscape of data security shifts in healthcare, Healthcare Information 33 (2016) 41−47.

[20] Y. Akao, Quality Function Deployment: Integrating Customer Requirements Into Product Design, Productivity Press, Cambridge, MA, 1990.

[21] J. Bossert, Quality Function Deployment — A Practitioner's Approach, ASQC Quality Press, Milwaukee WI, 1991.

[22] R. King, Designing Products and Services That Customer Wants, Productivity Press, Portland OR, 1995.

[23] L.K. Chan, M.L. Wu, Quality Function Deployment: a literature review, European Journal of Operations Research 143 (2002) 463—497 [9] J. R. Hauser, D. Clausing, The House of Quality, Harvard Business Review 66 (1988) 1—27.

[24] M. Barad, D. Gien, Linking improvement models to manufacturing strategies, International Journal of Production Research 39 (2001) 2675—2695.

[25] J. Pfeffer, Producing sustainable competitive advantage through the effective management of people, Academy Management Executive 9 (1995) 55—69.

[26] J.C. Flanagan, The critical incident technique, Psychological Bulletin 51 (1954) 327—358.

[27] N. Slack, The importance—performance matrix as a determinant of improvement priority, International Journal of Operations and Production Management 14 (1994) 59—75.

[28] C.M. Garcia, Strategies and performance in hospitals, Health Policy (Amsterdam, Netherlands) 67 (2004) 1—13.

[29] T.J. Douglas, Understanding competitive advantage in the general hospital industry: evaluating strategic competence, Strategic Management Journal 24 (2003) 334—347.

CHAPTER 5

Enhanced Galactic Swarm Algorithm with Encryption Technique for Medical Image Security in Internet of Things environment

S.P. Velmurugan[1], A.M. Gurusigaamani[2], P. Vigneshwaran[1], V. Suresh Babu[1] and Jenyfal Sampson[1]
[1]Department of Electronics and Communication Engineering, Kalasalingam Academy of Research and Education, Krishnankoil, Tamil Nadu, India
[2]Department of Computer Science Engineering, Kalasalingam Academy of Research and Education, Krishnankoil, Tamil Nadu, India

5.1 Introduction

The growth in the medical field and the development of medical devices like computed topography and magnetic resonance imaging generate large data every day, and the information gathered from such gadgets were rich in variables and highly dimensional [1,2]. Hence, the medical image datasets and their dimensionality are found to be raising enormously. Owing to this rise in medical databases, it becomes tough to manage the file system with expanding data volume [3]. Hence, managing medical data becomes the main concern for health-care service providers. Thus cloud computing (CC) has been used in the medicinal sector to compute and store health data as the medical image cloud was simple to manage [4,5]. CC renders scalable and flexible computational resources from remote areas and is accessible and reliant on the user's need. It is even effective to deliver computational resources in the high end—computing atmosphere [6]. Additionally, the Internet of Things (IoT) can be extended in CC to develop innovative applications and facilities in the medical process. In simple terms, IoT is the network of objects or things like electronic devices, sensors, and software connected to exchange data with manufacturers, operators, or other linked gadgets to attain greater services and values [7,8]. Likewise, IoT provides advanced connections among devices,

Securing Next-Generation Connected Healthcare Systems
DOI: https://doi.org/10.1016/B978-0-443-13951-2.00008-8

systems, and services involving different applications, domains, and protocols. CC and IoT equally profited when integrating technologies [9]. The IoT often supports the cloud to scale up the efficiency, namely, storage, computational ability, high resource, and energy utilization. Similarly, it favors the cloud to offer various new services over an active and distributed method [10].

The requirements to fulfill the security wants of digital imageries have provoked the enhancement of good encoding techniques [11]. For transmission and capacity, encryption was a very effective gadget, yet once the delicate information was decoded, the information was not secured any longer. Whenever the images were in open frame, the serious risk was the violation of the entering rights and the day-to-day logs by intruders [12,13]. The common encryption techniques put prominence on paired data or text data. Hence, the customary ciphers such as data encryption standard (DES), International Data Encryption Algorithm (IDEA), Rivest-Shamir-Adleman (RSA), and Advanced Encryption Standard (AES) were unsuitable for real-time image encryption since such ciphers need high registering power and an expansive computing period [14]. By considering the aids of encryption in providing security to exceptional data, providing a method will be effective for executing decryption and encryption on health images [15]. The ideologies of the optimization method are to decode and scramble images to interchange them between the receiver and the sender safely.

This chapter introduces an Enhanced Galactic Swarm Algorithm with an Encryption Technique for Medical Image Security (EGSAET-MIS) in the IoT environment. The presented EGSAET-MIS approach preprocesses the medical images using the median filtering (MF) technique. The EGSAET-MIS technique uses extended homomorphic encryption (EHE) techniques for medical image encryption. To enhance the security level of the EHE technique, the EGSAET-MIS technique applies the EGSA algorithm to choose the keys and optimizes the performance. In the presented EGSAET-MIS technique the maximum peak signal-to-noise ratio (PSNR) was considered the fitness function (FF). The simulation analysis of the EGSAET-MIS technique is examined on several medical images.

5.2 Related works

Elhoseny et al. [16] inspected the security of health-care images in IoT by using a new cryptographic technique with optimized algorithms.

Mostly, the patient data will be saved as a cloud server in the medical center since security is highly important. Therefore another structure can be needed to effectively store and secure the transmission of medical images interleaved with patient data. The optimal key was selected using hybrid swarm optimization, that is, particle swarm optimization (PSO) and grasshopper optimization in elliptic curve cryptography, to raise the security level of decryption and encryption. Khashan and AlShaikh [17] introduced a lightweight selective encryption method for encoding edge maps of health-care images. Through an edge detection technique the edge map was initially extracted. After that a chaotic map can be employed for generating large key spaces. The authors presented a onetime pad technique to encode the important identified image blocks.

Kamal et al. [18] modeled a new encryption approach to encrypt gray and color health images. An innovative image-splitting approach related to image blocks is presented. After, the image blocks scrambled through a random permutation, zigzag pattern, and rotation. Afterwards, the chaotic logistic map will generate a key for diffusing scrambled images. The performance of this modeled technique in encoding medical images was assessed through time complexity and security analysis. Lakshmi et al. [19] devise Hopfield neural networks (HNNs)-influenced encrypted image method to endure several assaults, which improvise and optimize the system by continual updating and learning. Such techniques present security feature that adapts themselves to everyday wonders of real time. This method used the back propagation neural network (BPNN) to generate image-specific keys that augmented resiliency against invaders. Using HNN, the generated keys will be employed as a primary seed for diffusion and confusion series generation.

In Ref. [20] the dual encryption procedure can be leveraged for encoding medical images. Primarily Blowfish Encryption was under consideration, and the signcryption method was leveraged to confirm the encrypted approach. Then, opposition-based flower pollination was employed to upgrade private and public keys (PKs). In Ref. [21], the Memetic Algorithm is an evolutionary method to encode text messages. The encoded data were fed into the medical images using DWT 1 and 2. In the extraction process of a hidden message from the encoded letter, the reverse algorithm of the Memetic Algorithm was adapted. Five Grayscale and five red green blue (RGB) images will be employed to test the proposed algorithm and show its accuracy.

5.3 The proposed model

This chapter has introduced a new EGSAET-MIS approach for medical image encryption in the IoT environment. In the presented EGSAET-MIS technique the major aim is to transmit medical images in the IoT platform securely. The presented EGSAET-MIS technique encompasses MF-based noise removal, EHE-based encryption, and EGSA optimal key generator. Fig. 5.1 represents the workflow of the EGSAET-MIS system.

5.3.1 Image preprocessing

The presented EGSAET-MIS technique encompasses MF-based noise removal process to boost the input image quality. Order-statistics filter, otherwise called MF, interchanges the prediction of a pixel by the middle gray levels in the region of that pixel [22]. The median was an intelligence

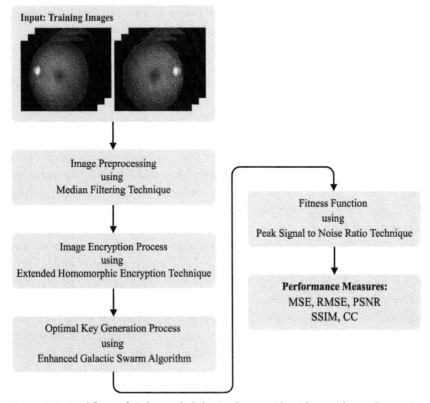

Figure 5.1 Workflow of Enhanced Galactic Swarm Algorithm with an Encryption Technique for Medical Image Security system.

and rank command statistic; the main stream of the pixel values will decide the outcome. The expression of MF is as follows:

$$f'(x, y) = \underset{(s,t)\in S_{xy}}{\text{median}} \{g(s, t)\} \tag{5.1}$$

The first pixel prediction was incorporated into the median calculation, which can help mitigate certain types of random noise. With stunningly less clouding than a linear smoothing filter of size, it renders astonishing noise diminish capabilities. The median was recognized by first sorting every value from the window numerically. A while later, displacing pixel deemed with the middle pixel value. MFs were feasible for unipolar and bipolar impulse noise. MF can be reasonable within the sight of both bipolar and unipolar impulse noise. Trimmed median noise removal filters had a potential role for images corrupted by scratches made by human beings, the atmosphere, etc.

5.3.2 Image encryption using extended homomorphic encryption technique

For medical image encryption the EGSAET-MIS technique uses the EHE technique. The homomorphic encryption (HE) method is an encryption technique that allows the conducting of mathematical operations on encrypted datasets [23]. Due to its high-performance capabilities in storing, processing, and transmitting encrypted information, which is implemented in numerous applications, namely the financial sector, healthcare, medical applications, social media advertisement, smart vehicles, and forensic applications, where maintaining the confidentiality of the user is of great significance. The HE scheme includes four stages, encryption, decryption, evaluation, and key generation.

Encryption: The user produces the ciphertext (C) with the help of PK and M and saves it on the server.

Decryption: The user attains (M) original text via decoding the evaluation through the cloud server through secret key (SK).

Evaluation: The server estimates C after the user receives the encoded results transmitted via the server.

Key generation: The user generates the SK, public parameter, and PK. In the inference stage:

- the storage and encryption of user information are performed within the cloud server,
- the user transmits data on the training process to the server, and

- the encrypted data through HE are fed into the server that transmits back the encrypted results to the user.

For the SK the customer could decrypt the results. Thereby, the security and privacy of the data are retained. Fig. 5.2 demonstrates the process of HE.

The EHE system was a HE branch that assists additive and multiplicative homomorphism with a constrained amount of processes. In 2010 the Dijk−Gentry−Halevi−Vaikuntanathan system was introduced as a second FHE system, an asymmetric cryptosystem system. However, it depends on the homomorphic properties with the small amount of operations. Thus it gives few properties of EHE. Different parameters are required for the execution of this system λ. Especially, η represents the bit-length of SK, γ indicates the bit-length of the integer in the PK, and ρ denotes the bit-length of the noise and number of integers in the PK. This scheme is defined in the succeeding three algorithms (Algorithm 5.1 [key generation], Algorithm 5.2 [encryption], and Algorithm 5.3 [decryption]).

5.3.3 Optimal key generation using EGSA

To improve the security level of the EHE technique, the EGSAET-MIS technique applies the EGSA algorithm to optimally choose the

Figure 5.2 Process in homomorphic encryption.

ALGORITHM 5.1 : SH.keyGenerate (λ).

Input: λ secret parameter

Output: private key: $sk = p$ and public key: $pk = (x_0, x_1, \ldots x_\tau)$

Randomly select the private key p, namely, odd number, whereas $p \in [2^{\eta-1}, 2^{\eta})$

and $\eta = \lambda^2$

Produce an array of integers, whereas $q_i \in Z, q_i \in [0, 2^\tau) q_i \neq p$, and $\gamma = \lambda^5$

Randomly choose $r_i \in Z$ and $r_i \in (-2^\rho, 2^\rho)$, whereas $\rho = 2\lambda$ and $i = 0, \ldots \tau$ with $\tau = \gamma + \lambda$

4: Determine the function: $x_i = pq_i + r_i$

5: x_0 large pk value and needs to be odd. Then, the remainder of x_0 needs to be even

ALGORITHM 5.2 : SH.Encryption (M, pk).

Input: M message to encode

Output: c encryption message

Consider a random subset $S \in (0, 1, \tau)$

Produce a random integer $r \in (-2^{\rho'}, 2^{\rho'})$

$$c = (M + 2r + 2\Sigma_{i \in S} x_i) \bmod x_0$$

ALGORITHM 5.3 : SH.Decryption (c, sk).

Input: c encrypt the message

Output: M original message

Evaluate $M = (c \bmod sk) \bmod 2$

keys and maximizes the performance. A novel GSA technique inspires the motion of galaxies and stars in the universe [24]. The galaxies and stars can be differently distributed in space. Next, the attraction of galaxies and stars from the GSA system was inspired. Primarily, the population initialized was classified as subpopulations termed sub-swarms; all the individuals of subswarms start their movement depen-dent upon the PSO technique by definite iteration amount, and all the individuals from all the subpopulations can be allured to separate with optimum fitness value; eventually, the iteration of subpopulations can be considered by the optimum different from all the subpopulations. An optimum individual of all the subpopulations permits the next step. In contrast, it can be a procedure novel superswarm, and GSA returns the optimum separation of the superswarm, distinguishing the opti-mum solution initiating in the entire population initialized. During the following pseudocode, all the necessary phases for executing the task of GSA are as follows:

- Considering the PSO system was employed as a beginning in GSA, it contains exploitation and exploration phases. In contrast, it utilizes parameters from the calculation of velocity that enhances or decreases as time proceeds to the exploitation step and permits controlling the exploration step.
- With the linear reduction the PSO technique inclines to have a global search capability but initiates the execution and local searching but approaches the end of execution.
- Such parameters can be acceleration constant that creates the solution method an optimum global as well as local solutions, and it can ini-tially proceed with constant value and alter the value in the mathemat-ical process.

In the dynamic, optimized issue, the environment alteration dynami-cally needs the searching procedure to search nonlinearly for the subse-quent altering environment. Therefore it requires dynamic alteration to an optimum balancing betwixt global as well as local searching (Algorithm 5.4).

Lens imaging is a physical optics phenomenon that signifies the detail that however an object has been placed at two or more rules of focal lengths beyond the convex lens, a lesser and reversed image can be cre-ated on the opposite side of the lens [25]. To create the 1D searching space, for example, there was a convex lens having focal length f fixed at base point O (the midpoint of searching range $[lb, ub]$). In addition, an object p with height h has located on the coordinate axis, and their pres-ent is GX (candidate solution). The distance in the object to lens u is superior to twice f. With lens imaging function a reversed imaging p' of height h^* can be obtained that is presented as GX^* (reverse solution) on

ALGORITHM 5.4 : Pseudocode of GSA algorithm

1. Initializing GSA.
2. The population has been classifier into M subpopulations.
    $X_i \subset X : i = 1, 2, \ldots, M$
3. The population was arbitrarily initialized
    $X_j^{(i)} \in X_i : j = 1, 2, \ldots, N$
4. Start Level1.
    Initiating PSO to all the M subpopulations and calculating the velocity and place of particles.
    $$v_j^{(i)} \leftarrow W_1 v^{(i)} + c_1 r_1 \left(p_j^{(i)} - x_j^{(i)} \right) + c_2 r_2 \left(g^{(i)} - x_j^{(i)} \right)$$
    $$x_j^{(i)} \leftarrow x_j^{(i)} + v_j^{(i)}$$
    Stop PSO.
    Stop Level1.
5. Start Level2.
    Initializing the superswarm.
    $Y^{(i)} \in Y : i = 1, 2, \ldots, M$
    Start PSO.
    Calculate the velocity and location of particles.
    $$v^{(i)} \leftarrow W_2 v^{(i)} + c_3 r_3 \left(p^{(i)} - Y^{(i)} \right) + c_4 r_4 \left(g - Y^{(i)} \right)$$
    $$Y^{(i)} \leftarrow Y^{(i)} + v^{(i)}$$
    Stop PSO.
    Stop Level2.
6. Return the optimum place g and fitness value $f(g)$.
7. Stop GSA.

the x-axis. According to the procedures of lens imaging and the same triangle, the geometrical connection reached is formulated as:

$$\frac{(lb + ub)/2 - GX}{GX^* - (lb + ub)/2} = \frac{h}{h^*} \tag{5.2}$$

At this point, assume the scale factor $n = h/h^*$, the inverse solution GX^* was computed as transmitting:

$$GX^* = \frac{lb + ub}{2} + \frac{lb + ub}{2n} - \frac{GX}{n} \tag{5.3}$$

It can be noticeable that once $n = 1$, it has simplified as the common equation of the OBL approach:

$$GX^* = lb + ub - GX \tag{5.4}$$

Therefore it can assume that the OBL method is a peculiar case of LOBL. Compared with OBL, the latter permits attaining dynamic inverse solution and extensive searching range by tuning the scale feature n.

It can usually be extended to D-dimension space:

$$GX_j^* = \frac{lb_j + ub_j}{2} + \frac{lb_j + ub_j}{2n} - \frac{GX_j}{n} \tag{5.5}$$

where lb_j and ub_j represent the lower and upper limits of jth dimensional correspondingly, $j = 1, 2, \cdots D$, GX_j^* refers to the reverse solution of GX_j from the jth dimensional.

If the novel inverse solution was created, there is no guarantee that it can be continuously superior to the present candidate solution from the gorilla place. So, it can be essential for evaluating the fitness values of inverse as well as candidate solutions, next the fitter one is chosen for continuing participate from the subsequent exploitation step which is defined as:

$$GX_{next} = \begin{cases} GX^*, & \text{if } F(GX^*) < F(GX) \\ GX, & \text{otherwise} \end{cases} \tag{5.6}$$

where GX^* refers to the inverse solution created by LOBL, GX implies the present candidate solution, GX_{ext} denotes the chosen gorilla for continuing the following place upgrade, and F implies the FF of problems.

An optimal key selective technique considered that "fitness function" can be maximal key utilizing PSNR for unscrambling and scrambling data in medical images in IoT. The process was produced by a hybrid optimized method for evaluation. It is illustrated in the following condition:

$$\text{fitness} = \text{maximum}\{\text{PSNR}\} \tag{5.7}$$

5.4 Experimental validation

The experimental validation of the EGSAET-MIS method is tested using a set of medical images. Fig. 5.3 showcases the sample images.

Brief encryption results of the EGSAET-MIS model under different test images in terms of correlation coefficeint (CC) and structural similarity (SSIM) are illustrated in Table 5.1 and Fig. 5.4. The outcomes inferred that the EGSAET-MIS method had reached effectual outcomes under all images. For example, with image-1, the EGSAET-MIS method has rendered CC and SSIM 99.96% and 99.95%, respectively. Similarly, with

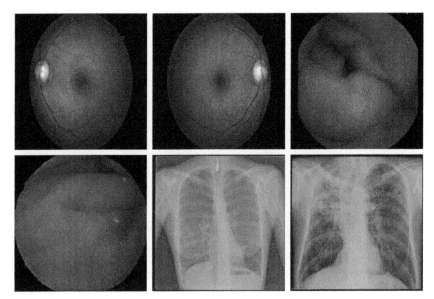

Figure 5.3 Sample images.

Table 5.1 Result analysis of Enhanced Galactic Swarm Algorithm with an Encryption Technique for Medical Image Security (EGSAET-MIS) system with distinct measures and images.

Test images	MSE	RMSE	PSNR	SSIM	CC
Image1	0.0816	0.2857	59.01	99.95	99.96
Image2	0.0810	0.2846	59.05	99.96	99.91
Image3	0.0706	0.2657	59.64	99.91	99.95
Image4	0.0756	0.2750	59.35	99.96	99.94
Image5	0.0813	0.2851	59.03	99.94	99.93
Image6	0.0737	0.2715	59.46	99.90	99.94

CC, Cloud computing; *PSNR*, peak signal-to-noise ratio.

image-2, the EGSAET-MIS method has offered CC and SSIM of 99.91% and 99.96% correspondingly. In addition, with image-4, the EGSAET-MIS approach has rendered CC and SSIM of 99.94% and 99.96% correspondingly. Meanwhile, with image-5, the EGSAET-MIS methodology has offered CC and SSIM of 99.93% and 99.94% correspondingly. Eventually, with image-6, the EGSAET-MIS approach has offered CC and SSIM of 99.94% and 99.90% correspondingly.

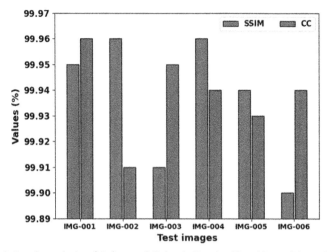

Figure 5.4 Result analysis of Enhanced Galactic Swarm Algorithm with an Encryption Technique for Medical Image Security system with distinct measures and images.

Table 5.2 MSE analysis of Enhanced Galactic Swarm Algorithm with an Encryption Technique for Medical Image Security (EGSAET-MIS) system with other approaches under several test images.

	MSE					
Test images	EGSAET-MIS	PIOESMIM	OSCMI	Hybrid	GWOESCO	SCECC
IMG_001	0.0816	0.1038	0.1240	0.1435	0.1704	0.1952
IMG_002	0.0810	0.0964	0.1206	0.1376	0.1542	0.1689
IMG_003	0.0706	0.0884	0.1012	0.1220	0.1434	0.1741
IMG_004	0.0756	0.0955	0.1190	0.1421	0.1558	0.1729
IMG_005	0.0813	0.1013	0.1113	0.1215	0.1322	0.1568
IMG_006	0.0737	0.0861	0.0979	0.1127	0.1245	0.1377

The mean square error (MSE) analysis of the EGSAET–MIS method with recent approaches under several test images is shown in Table 5.2 and Fig. 5.5 [16,20,26]. The gained values implied that the EGSAET–MIS method had improved performance with the least values of MSE. For example, on IMG_001, the EGSAET–MIS methodology has reached a minimal MSE of 0.0816, while the PIOESMIM, OSCMI, hybrid, GWOESCO, and SCECC approaches have reported increased MSE of 0.1038, 0.1240, 0.1435, 0.1704, and 0.1952 correspondingly. In addition, on IMG_003, the

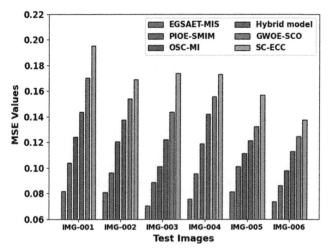

Figure 5.5 MSE analysis of Enhanced Galactic Swarm Algorithm with an Encryption Technique for Medical Image Security system under several test images.

EGSAET-MIS approach has attained a minimal MSE of 0.0706, while the PIOESMIM, OSCMI, hybrid, GWOESCO, and SCECC techniques have reported increased MSE of 0.0884, 0.1012, 0.1220, 0.1434, and 0.1741 correspondingly. Moreover, on IMG_006, the EGSAET-MIS approach has reached a minimal MSE of 0.0737, while the PIOESMIM, OSCMI, hybrid, GWOESCO, and SCECC methods have reported increased MSE of 0.0861, 0.0979, 0.1127, 0.1245, and 0.1377 correspondingly.

The comparative PSNR investigation of the EGSAET-MIS method with recent methods takes place in Table 5.3 and Fig. 5.6. The obtained values indicated that the EGSAET-MIS approach has superior performance with the least PSNR values. For instance, on IMG_001, the EGSAET-MIS model has demonstrated an increased PSNR of 59.01 dB, while the PIOESMIM, OSCMI, hybrid, GWOESCO, and SCECC models have revealed reduced PSNR of 57.97, 57.20, 56.56, 55.82, and 55.23dB, respectively. Additionally, on IMG_004, the EGSAET-MIS approach has demonstrated an increased PSNR of 59.35 dB, while the PIOESMIM, OSCMI, hybrid, GWOESCO, and SCECC approaches have revealed reduced PSNR of 58.33, 57.38, 56.60, 56.21, and 55.75 dB correspondingly. Eventually, on IMG_006, the EGSAET-MIS technique has shown an increased PSNR of 59.46 dB, while the PIOESMIM, OSCMI, hybrid, GWOESCO, and SCECC models have revealed reduced PSNR of 58.78, 58.22, 57.61, 57.18, and 56.74 dB correspondingly.

Table 5.3 Peak signal-to-noise ratio (PSNR) analysis of Enhanced Galactic Swarm Algorithm with an Encryption Technique for Medical Image Security (EGSAET-MIS) system with other approaches under several test images.

	PSNR (dB)					
Test Images	EGSAET-MIS	PIOESMIM	OSCMI	Hybrid	GWOESCO	SCECC
IMG_001	59.01	57.97	57.20	56.56	55.82	55.23
IMG_002	59.05	58.29	57.32	56.74	56.25	55.85
IMG_003	59.64	58.67	58.08	57.27	56.57	55.72
IMG_004	59.35	58.33	57.38	56.60	56.21	55.75
IMG_005	59.03	58.07	57.67	57.29	56.92	56.18
IMG_006	59.46	58.78	58.22	57.61	57.18	56.74

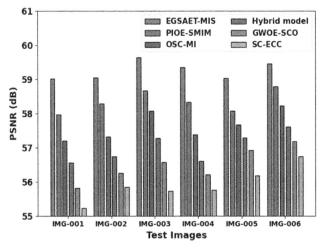

Figure 5.6 Peak signal-to-noise ratio analysis of Enhanced Galactic Swarm Algorithm with an Encryption Technique for Medical Image Security system under several test images.

The comparative CC study of the EGSAET–MIS technique with current approaches takes place in Table 5.4 and Fig. 5.7. The gained values show that the EGSAET–MIS method has accomplished superior performance with the least CC values. For example, on IMG_001, the EGSAET–MIS approach has demonstrated increased CC of 99.96, while the PIOESMIM, OSCMI, hybrid, GWOESCO, and SCECC models have revealed reduced CC of 98.38, 97.11, 96.13, 95.40, and 94.11 correspondingly. Moreover, on IMG_004, the EGSAET–MIS model has demonstrated increased CC of

Table 5.4 Cloud computing analysis of Enhanced Galactic Swarm Algorithm with an Encryption Technique for Medical Image Security (EGSAET-MIS) system with other techniques under several test images.

Test images	EGSAET-MIS	PIOESMIM	OSCMI	Hybrid	GWOESCO	SCECC
			Correlation coefficient			
IMG_001	99.96	98.38	97.11	96.13	95.40	94.11
IMG_002	99.91	98.48	97.30	95.98	95.30	93.62
IMG_003	99.95	98.87	97.64	96.56	95.07	94.43
IMG_004	99.94	99.32	98.57	97.92	96.82	94.98
IMG_005	99.93	98.80	97.56	96.31	95.38	94.12
IMG_006	99.94	98.38	97.52	96.56	95.42	95.23

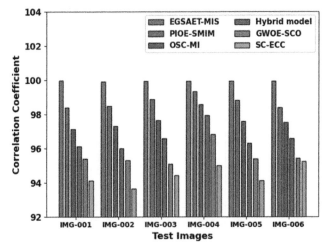

Figure 5.7 Cloud computing analysis of Enhanced Galactic Swarm Algorithm with an Encryption Technique for Medical Image Security system under several test images.

99.94, while the PIOESMIM, OSCMI, hybrid, GWOESCO, and SCECC methods have revealed reduced CC of 99.32, 98.57, 97.92, 96.82, and 94.98 correspondingly. Eventually, on IMG_006, the EGSAET-MIS method exhibited increased CC of 99.94, while the PIOESMIM, OSCMI, hybrid, GWOESCO, and SCECC methods have revealed reduced CC of 98.38, 97.52, 96.56, 95.42, and 95.23 correspondingly.

The detailed PSNR examination of the EGSAET-MIS method with recent approaches takes place with the attack in Table 5.5. The obtained

Table 5.5 Comparative peak signal-to-noise ratio (PSNR) analysis of Enhanced Galactic Swarm Algorithm with an Encryption Technique for Medical Image Security (EGSAET-MIS) system with other techniques under several test images.

Methods	Attack Type	PSNR (dB)					
		IMG_001	IMG_002	IMG_003	IMG_004	IMG_005	IMG_006
EGSAET-MIS	With attack	59.73	58.61	58.33	58.54	58.38	58.10
	WOA	60.83	59.80	59.45	59.38	59.76	59.37
PIOESMIM	With attack	58.30	56.71	56.60	56.87	56.33	56.41
	WOA	59.22	57.91	57.61	58.15	57.34	57.65
OSCMI	With attack	56.47	54.61	54.51	55.45	54.86	54.92
	WOA	57.45	55.61	55.85	56.58	56.01	55.75
Hybrid	With attack	55.00	52.92	52.77	54.05	53.31	53.09
	WOA	56.35	54.24	53.59	55.37	54.38	54.22
GWOESCO	With attack	52.94	51.14	51.37	52.15	51.59	51.68
	WOA	54.19	52.45	51.94	53.44	52.42	51.80
SCECC	With attack	50.11	50.96	50.31	50.95	50.16	50.77
	WOA	51.27	51.81	51.26	52.21	51.15	52.17

values show that the EGSAET-MIS method has accomplished superior performance with the least PSNR values. For instance, on IMG_001 with the attack, the EGSAET-MIS method has exhibited an increased PSNR of 59.73 dB, while the PIOESMIM, OSCMI, hybrid, GWOESCO, and SCECC models have revealed reduced PSNR of 58.30, 56.47, 55, 52.94, and 50.11 dB correspondingly. Concurrently, on IMG_003 with the attack, the EGSAET-MIS method has shown an increased PSNR of 58.33 dB, while the PIOESMIM, OSCMI, hybrid, GWOESCO, and SCECC approaches have revealed a reduced PSNR of 56.60, 54.51, 52.77, 51.37, and 50.31 dB correspondingly. Also, on IMG_006 with the attack, the EGSAET-MIS technique has demonstrated an increased PSNR of 58.10 dB, while the PIOESMIM, OSCMI, hybrid, GWOESCO, and SCECC methods have revealed reduced PSNR of 56.41, 54.92, 53.09, 51.68, and 50.77 dB correspondingly.

The acquired values show that the EGSAET-MIS method has accomplished superior performance with least PSNR values. For example, on IMG_001 without attack, the EGSAET-MIS method has demonstrated an increased PSNR of 60.83 dB, while the PIOESMIM, OSCMI, hybrid, GWOESCO, and SCECC approaches have revealed reduced PSNR of 59.22, 57.45, 56.35, 54.19, and 51.27 dB correspondingly. In parallel, on IMG_003 without attack, the EGSAET-MIS approach has demonstrated an increased PSNR of 59.45 dB, while the PIOESMIM, OSCMI, hybrid, GWOESCO, and SCECC methods have revealed reduced PSNR of 57.61, 55.85, 53.59, 51.94, and 51.26 dB correspondingly. Also, on IMG_006 without attack, the EGSAET-MIS approach has exhibited an increased PSNR of 59.37 dB, while the PIOESMIM, OSCMI, hybrid, GWOESCO, and SCECC methods have revealed reduced PSNR of 57.65, 55.75, 54.22, 51.80, and 52.17 dB correspondingly. These results highlighted the supremacy of the EGSAET-MIS model on medical image security in the IoT platform.

5.5 Conclusion

This chapter has introduced a new EGSAET-MIS technique for medical image encryption in the IoT environment. In the presented EGSAET-MIS technique, the major aim is to transfer medical images in the IoT platform securely. The presented EGSAET-MIS technique encompasses MF-based noise removal, EHE-based encryption, and EGSA optimal key generator. To improvize the security level of the EHE technique, the

EGSAET-MIS technique applies the EGSA algorithm to optimally choose the keys and maximizes the performance. In the presented EGSAET-MIS technique, the increase in PSNR values is considered the FF. The simulation analysis of the EGSAET-MIS technique is examined on several medical images. The simulation outcomes depicted that the EGSAET-MIS method surpasses the other existing encryption techniques in terms of different measures.

Conflict of interest

The authors declare that they have no conflict of interest. The manuscript was written through contributions of all authors. All authors have given approval to the final version of the manuscript.

Data availability statement

Data sharing is not applicable to this chapter as no datasets were generated during the current study.

Ethics approval

This chapter does not contain any studies with human participants performed by any of the authors.

Consent to participate

Not applicable.

Informed consent

Not applicable.

References

[1] M. Sultana, A. Hossain, F. Laila, K.A. Taher, M.N. Islam, Towards developing a secure medical image sharing system based on zero trust principles and blockchain technology, BMC Medical Informatics and Decision Making 20 (1) (2020) 1−10.

[2] S. Namasudra, A secure cryptosystem using DNA cryptography and DNA steganography for the cloud-based IoT infrastructure, Computers and Electrical Engineering 104 (2022) 108426. Available from: https://doi.org/10.1016/j.compeleceng.2022.108426.

[3] O. Hosam, M.H. Ahmad, Hybrid design for cloud data security using combination of AES, ECC and LSB steganography, International Journal of Computer Sciences and Engineering 19 (2) (2019) 153–161.

[4] K.C. Nunna, R. Marapareddy, Secure data transfer through internet using cryptography and image steganography, SoutheastCon 2 (2020) 1–5. Available from: https://doi.org/10.1109/SoutheastCon44009.2020.9368301.

[5] L. Kumar, A secure communication with one time pad encryption and steganography method in cloud, Turkish Journal of Computer and Mathematics Education (TURCOMAT) 12 (10) (2021) 2567–2576.

[6] W.A. Awadh, A.S. Hashim, A. Hamoud, A review of various steganography techniques in cloud computing, University of Thi-Qar Journal of Science 7 (1) (2019) 113–119.

[7] M.S. Abbas, S.S. Mahdi, S.A. Hussien, Security improvement of cloud data using hybrid cryptography and steganography, in: International Conference on Computer Science and Software Engineering (CSASE), 2020, pp. 123–127. Available from: https://doi.org/10.1109/CSASE48920.2020.9142072.

[8] B. Abd-El-Atty, A.M. Iliyasu, H. Alaskar, A.A.Abd El-Latif, A robust quasi-quantum walks-based steganography protocol for secure transmission of images on cloud-based E-healthcare platforms, Sensors 20 (11) (2020) 3108.

[9] M.O. Rahman, M.K. Hossen, M.G. Morsad, A. Chandra, An approach for enhancing security of cloud data using cryptography and steganography with e-lsb encoding, IJCSNS 18 (9) (2018) 85.

[10] S. Arunkumar, S. Vairavasundaram, K.S. Ravichandran, L. Ravi, RIWT and QR factorization based hybrid robust image steganography using block selection algorithm for IoT devices, Journal of Intelligent & Fuzzy Systems 36 (5) (2019) 4265–4276.

[11] R. Adee, H. Mouratidis, A dynamic four-step data security model for data in cloud computing based on cryptography and steganography, Sensors 22 (3) (2022) 1109.

[12] K.B. Madavi, P.V. Karthick, Enhanced cloud security using cryptography and steganography techniques, in: International Conference on Disruptive Technologies for Multi-Disciplinary Research and Applications (CENTCON), vol. 1, 2021, pp. 90–95. Available from: https://doi.org/10.1109/CENTCON52345.2021.9687919.

[13] A. Sukumar, V. Subramaniyaswamy, V. Vijayakumar, L. Ravi, A secure multimedia steganography scheme using hybrid transform and support vector machine for cloud-based storage, Multimedia Tools and Applications 79 (15) (2020) 10825–10849.

[14] M.A. Khan, Information security for cloud using image steganography, Lahore Garrison University Research Journal of Computer Science and Information Technology 5 (1) (2021) 9–14.

[15] A.A. AB, A. Gupta, S. Ganapathy, A new security mechanism for secured communications using steganography and CBA, ECTI Transactions on Computer and Information Technology (ECTI-CIT) 16 (4) (2022) 460–468.

[16] M. Elhoseny, K. Shankar, S.K. Lakshmanaprabu, A. Maseleno, N. Arunkumar, Hybrid optimization with cryptography encryption for medical image security in Internet of Things, Neural Computing and Applications 32 (15) (2020) 10979–10993.

[17] O.A. Khashan, M. AlShaikh, Edge-based lightweight selective encryption scheme for digital medical images, Multimedia Tools and Applications 79 (35) (2020) 26369–26388.

[18] S.T. Kamal, K.M. Hosny, T.M. Elgindy, M.M. Darwish, M.M. Fouda, A new image encryption algorithm for grey and color medical images, IEEE Access 9 (2021) 37855–37865. Available from: https://doi.org/10.1109/ACCESS.2021.3063237.

[19] C. Lakshmi, K. Thenmozhi, J.B.B. Rayappan, S. Rajagopalan, R. Amirtharajan, et al., Neural-assisted image-dependent encryption scheme for medical image cloud storage, Neural Computing and Applications 33 (12) (2021) 6671–6684.

[20] T. Avudaiappan, R. Balasubramanian, S.S. Pandiyan, M. Saravanan, S.K. Lakshmanaprabu, et al., Medical image security using dual encryption with oppositional based optimization algorithm, Journal of Medical Systems 42 (11) (2018) 1–11.

[21] S. Doss, J. Paranthaman, S. Gopalakrishnan, A. Duraisamy, S. Pal, et al., Memetic optimization with cryptographic encryption for secure medical data transmission in IoT-based distributed systems, Computers, Materials & Continua 66 (2) (2021) 1577–1594.

[22] H.M. Ali, MRI medical image denoising by fundamental filters, High-Resolution Neuroimaging-Basic Physical Principle., Clinical Applications 14 (2018) 111–124. Available from: https://doi.org/10.5772/intechopen.72427.

[23] W. Boulila, M.K. Khlifi, A. Ammar, A. Koubaa, B. Benjdira, et al., A hybrid privacy-preserving deep learning approach for object classification in very high-resolution satellite images, Remote Sensing 14 (18) (2022) 4631.

[24] S. Bhardwaj, G.B.D. Amali, A. Phadke, K.S. Umadevi, P. Balakrishnan, A new parallel galactic swarm optimization algorithm for training artificial neural networks, Journal of Intelligent & Fuzzy Systems 38 (5) (2020) 6691–6701.

[25] Q. Fan, Z. Chen, W. Zhang, X. Fang, ESSAWOA: enhanced whale optimization algorithm integrated with salp swarm algorithm for global optimization, Engineering with Computers (2020) 1–18. Available from: https://doi.org/10.1007/s00366-020-01189-3.

[26] B.T. Geetha, P. Mohan, A.V. Mayuri, T. Jackulin, J.L. Aldo Stalin, et al., Pigeon inspired optimization with encryption based secure medical image management system, Computational Intelligence and Neuroscience 2022 (2022) 1–13. Available from: https://doi.org/10.1155/2022/2243827.

CHAPTER 6

An efficient heart disease prediction model using particle swarm—optimized ensemble classifier model

Priyanka Dhaka[1], Ruchi Sehrawat[2] and Priyanka Bhutani[2]
[1]University School of Information and Communication Technology and Maharaja Surajmal Institute, GGSIPU, Delhi, India
[2]University School of Information and Communication Technology, GGSIPU, Delhi, India

6.1 Introduction

Health is not only the lack of disease, but it is also totally a mental, well-being, and physical state. People's desire for a better life includes a vital aspect of health. Due to a variety of problems, such as subpar health-care services, significant differences between urban and rural areas, and a scarcity of doctors and nurses during the most challenging periods, the global health issue has unfortunately produced a conundrum [1].

Health conditions can be diagnosed diagonally and prevented in their early stages with the right care. With the use of various diagnostic tools like CT, MRI, and PET, abnormalities that exist within our bodies might be discovered rapidly. The unpredictability of the development of progressive illnesses has been a health-care system problem, and there is a strong need for funds for the whole thing from hospital beds to doctors and nurses as a result of the world's enormous population growth [2]. One of today's most essential and difficult health issues is the automatic detection of heart disease [3].

Early disease detection and diagnosis are made possible with remote patient monitoring, and the medical history of the patient can be saved in a database for future exploit. Since medical servers and databases contain medical records, they are essential for getting the relevant patient's health records [4]. A wide number of health-care applications are projected to benefit from Internet of Things (IoT) and wearable monitoring devices. The IoT was quickly adopted by the health-care sector [5,6] because

Securing Next-Generation Connected Healthcare Systems
DOI: https://doi.org/10.1016/B978-0-443-13951-2.00005-2

incorporating IoT features into medical devices improves both the quality and efficiency of services. A heart disease monitoring device that uses IoT technology may record and transfer the patient's physical information in real time to a hospital center located far away [7]. The human body's most vital organ is the heart. Heart disease is a serious health problem [8]. Data mining and machine learning (ML) approaches shorten computation times and costs. One of the applications of ML is the identification of medical conditions and interventions to improve a patient's daily life. Heart disease is typically thought to predominantly afflict the elderly, although it is actually spreading more widely among all age groups.

* *Ensemble classifier model*: The ensemble classifier model is developed by the standard hybridization of deep convolution neural network (CNN) and the BiLSTM which effectively learns the features and makes the predictions more accurate.
* *Partical swarm optimization (PSO)-optimized ensemble classifier model*: PSO optimization effectively tunes the parameters by increasing the velocity which minimizes the computational time and provides the optimized output.
* The optimization tuned the ensemble classifier effectively and makes the predictions with less computational complexity.

The chapter is written as follows in accordance with its organizational structure: existing efforts and their methodology are described in Section 6.2. The system model for heart disease detection is explained in Section 6.3. The interpretation of the particle swarm optimization is in Section 6.4. In part 5 the results of heart disease detection are provided, in Section 6.6, the conclusion is provided.

6.2 Literature review

An Internet of Thing based prediction model utilized the MDCNN classifier which was introduced by Khan [3]. Although this method produced better performance and accuracy, it was unable to encode object location and orientation. An ensemble deep learning technique was reported by Ali et al. [9]. This technique produced better dimensionality and accuracy; however, it has an over fitting problem. A decision tree model was created by Basheer et al. [4] for the use of a continuous and patient monitoring remote system in the early heart disease prediction process. Although this approach was simple to comprehend and implemented, it was too slow to be effective. A smart health-care system for IoT was presented by

Islam et al. [1] that can track a patient's vital signs in real time. Although it was quite expensive, this approach produced high sensitivity and quick response times. An effective neural network with convolutional layers was created by Dutta et al. [1] to categorize clinical data that was highly class-imbalanced. The test accuracy of this technique was higher, but the classifiers had poor recall values, which is the actual positive rate for detecting coronary heart disease. Reddy et al. [10] presented a correlation-based feature selection method which was presented for feature extraction that detected coronary heart disease and three single classifiers and three ensemble classifiers were used for selecting the features that performed better and went through the overfitting problem. Taylam et al. [11] has explained the adaptive neuro-fuzzy and statistical approach naïve bays classifier for effective classification.

6.2.1 Challenges

- A difficult task for heart disease prediction is extracting pertinent and significant information from the data.
- Another difficulty for ML-based systems is choosing useful feature weights after selecting features from structured data.
- Partitioning a severely unbalanced dataset presents many difficulties and introduces biases that cannot be avoided in the performance of a classifier.

6.3 System model of smart healthcare

The PSO-optimized ensemble classifier system is composed of three operational phases: accurate disease prediction utilizing the acquired data, device and patient identification using an encryption method on a modified ECC Diffie–Huffman (EDH) algorithm, and data collection using dispersed IoT nodes.

To provide health status prediction, IoT nodes scattered across the sensing environment to acquire patient data in step one, as shown in Fig. 6.1. The advancement of encryption-based cloud data protection ensures that the data are safe from hackers after they have been collected. As a result, the data are securely sent to the cloud server, where the clinician would obtain the patient's medical records and examine the information to offer a diagnosis suggestion.

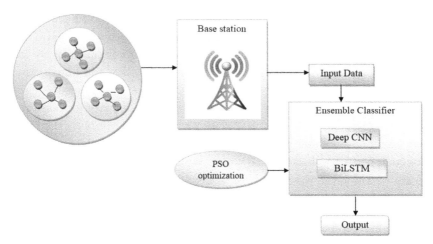

Figure 6.1 Diagrammatic representation of heart disease prediction.

6.3.1 Data acquisition and authentication

Data acquisition is the process of compiling information from all pertinent sources. It is essential to protect the patient information gathered during the monitoring phase since it often helps clinicians to identify patients correctly. To prevent unwanted users from accessing the data stored in the cloud, the collected data must be encoded. For the purpose of securely updating patient medical records on the cloud server, throughout the authentication process, the information is verified and approved. The previously published and current modified EDH is used to encrypt medical reports. This technique generates a key, produces a signature at the sender end, and validates the signature at the receiver end. The acquired data are then securely transferred to the target via a modified EDH, which the authorized clinician decrypts to offer the patients a diagnosis.

6.4 Ensemble classifier model for disease prediction

To train the preprocessed heart disease data, the deep CNN and BiLSTM are combined in the ensemble classifier. Due to the combined performance, ensemble classifier model achieves superior accuracy and efficiency against the deep learning model.

6.4.1 Deep CNN model

Heart disease is predicted using the deep CNN. It is a basic feed-forward neural network that has been trained using a PSO-optimized ensemble classifier. Fig. 6.2 illustrates the representation of CNN Architecture. It is used to categorize the data and automatically identify the important features. Four layers are presented in the deep CNN, and they are outlined next.

1. Convolutional layer:

   A convolutional layer produces several feature maps by extracting a range of local patterns at each small location in the input space using a set of convolutional filters.

2. Pooling layer:

   The pooling layer creates a single output from local regions of the convolution feature maps to downsample the feature maps produced from the earlier convolutional layers.

3. Fully connected layer

   The outcomes from earlier layers in this layer are integrated to provide the activation function that is a sigmoid function, depending on whether the final feature representations are for classification or regression (Figs. 6.2 and 6.3).

6.4.2 BiLSTM

A brand-new technique called the BiLSTM has emerged in recent years. The heart disease and Kaggle database are subjected to both forward and backward analyses in this method. A Bi-LSTM has three layers. Input, forget, and output gates are all there in the memory block that is located in the hidden layers. These gates regulate both the features and the amount of the extracted data. Depending on the forget and the input gates, the most recent information will be stored in a cell state, also the outdated data will be removed. The output gate completes the output of

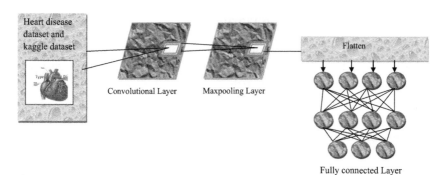

Figure 6.2 CNN architecture.

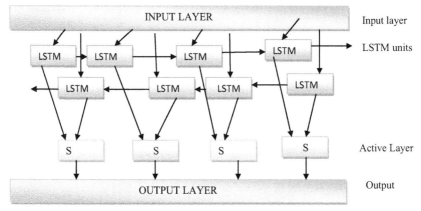

Figure 6.3 Diagrammatic representation of BiLSTM.

the network model after the end of the cell state update. This makes it possible for the output of Bi-LSTM units to generate a representation that considered both past and future data automatically.

$$HF = u(e_1 y + e_2 H_{F-1} + S_a) \tag{6.1}$$

$$HB = u(e_3 y + e_5 H_{B+1} + S_b) \tag{6.2}$$

$$HO = h(e_4 H + e_6 H_B + S_c) \tag{6.3}$$

HF denotes the forward layer, HB denotes the backward layer, HO denotes the output layer, u is an update function for the cells' state vectors in the Bi-LSTM block's hidden layers, h denotes the output layer activation function, weight is denoted as e, and bias is denoted as S.

6.5 Particle swarm optimization

Position and velocity are the main characteristics of the PSO algorithm's particles. To determine the value of the objective function for each particle position, a population of particles is randomly placed anywhere inside the search region. The initial values of the velocities are set to zero or chosen at random inside predetermined bounds. Therefore if U_j are the velocities of the j particles, then a starting setting is possible as follows:

$$U_j = 0j = 1, 2, \ldots, m \tag{6.4}$$

where m denotes the population size.

The initial velocities can be assigned in a variety of ways. The velocity of each particle is then updated using the proper formula. These velocity measurements are used to update the particle positions. After that it is decided if the objective function has improved. The values of the particle positions are considered the new positions if improvements are made. Additionally, the swarm's global optimal position for each particle is updated. Until a specified convergence condition is fulfilled, this process is repeated. After that the process is terminated and the answers are given. Then provide the precise algorithm for updating the velocity and position values.

The following equation, applied at time s, yields the velocity of each particle. In reality, the values of s correspond to the process's iteration or generation numbers, starting at $s = 0$.

$$U_j^{(s+1)} = U_j^{(s)} + dp_1 \cdot \left[Y_{\text{pbest}}^{(s)} - Y_j^{(s)} \right] + cp_2 \cdot \left[Y_{\text{gbest}}^{(s)} - Y_j^{(s)} \right] \qquad (6.5)$$

where d and c are typically positive constants that the user sets at the process's beginning. These parameters need to be carefully chosen because they have a big impact on the process's effectiveness and how the technique converges, properly emphasizing the local and global search components of the algorithm. $Y_{\text{pbest}}^{(s)}$ represents the particle's optimal location vector in the immediate area as determined by the objective function $f(Y_j)$. $Y_{\text{gbest}}^{(s)}$ represents the current best position in the world. This vector provides the ultimate optimal position vector and is continuously updated during each cycle. The current values of the velocity and position vectors are $U_j^{(s)}$ and $Y_j^{(s)}$, respectively. At each stage of the method, the values p_1, p_2 are reselected from the uniform random distribution vector p in the range of $0-1$. $\cdot$ in Eq. (6.5) stands for the Hadamard or Schur multiplication. For instance, if c and d are two-element vectors, then the equation becomes

$$d \cdot c = \begin{bmatrix} d_1 \\ d_2 \end{bmatrix} \cdot \begin{bmatrix} c_1 \\ c_2 \end{bmatrix} = \begin{bmatrix} d_1 & c_1 \\ d_2 & c_2 \end{bmatrix} \qquad (6.6)$$

The velocity adjustment equation's final two terms, Eq. (6.5), which indicate the algorithm's two opposing aspects, must be emphasized. First, make sure that the local area has been sufficiently explored to determine an acceptable level of local minimum accuracy. To find a global minimum and prevent being stuck at a local minimum, it is also important to make sure that the entire region is investigated. In this procedure, randomness is important. Therefore the selection of d and c in Eq. (6.2) is challenging and essential in assuring the compatibility of these two goals. The updated velocity values can be used to calculate the new position values.

$$Y_j^{(s+1)} = Y_i^{(s)} + U_j^{(s+1)} \tag{6.7}$$

This equation drives the process along to the next point, where it can be improved after being tested. Large values of velocity indicate quick exploration of the region but possible omission of important optima. The algorithm may advance very slowly if modifications to Y_j are made inanely slowly. A thoughtful selection of parameters can guarantee steady advancement.

6.6 Result

This section illustrates the effectiveness of the ensemble classifier model and other traditional methods for predicting heart disease using the Kaggle [7] and heart disease dataset [12].

6.6.1 Experimental setup

Utilizing the MATLAB® program in Windows 10 with 8-GB RAM, an ensemble classifier model is implemented.

6.6.2 Dataset description

Kaggle dataset (DB1) [12]: This database is produced by integrating a number of Hungary, Cleveland, and statlog datasets that are independently available and that meet specific criteria. The Cleveland database has 303 instances of heart, but the Hungarian database only has 270.

Heart disease dataset (DB2) [13]: This database offers 76 attributes and only 14 datasets are employed in the research. These data are used in a preliminary study to establish the absence or presence of heart disease.

6.6.3 Comparative methods

Hybrid Fuzzy—based DT [13] (T-1), HuS-based deep CNN (T-2), Hybrid social hunting—based deep CNN (T-3), deep BiLSTM [14] (T-4), and WOM-based deep BiLSTM (T-5) are the methods.

6.6.3.1 Comparative based on the DB1

The performances of the existing ensemble classifier and the PSO-optimized ensemble classifier are compared in Fig. 6.4. First, the suggested PSO-optimized ensemble classifier's accuracy improvement rate is measured and it achieved a rate of 0.10% improvement when compared to the WOM-based deep BiLSTM in Fig. 6.4A. The PSO–optimized ensemble classifier, which achieved a rate of 0.10% improvement when compared to the prior technique, is also measured to determine the improvement rate in terms of F–measure as shown in Fig. 6.4B. Similarly, Fig. 6.4C shows that the sensitivity is evaluated and an improvement rate of 0.05% is obtained. When comparing the final results to the previous method, the specificity measurement showed an improvement rate of 0.15% in Fig. 6.4D.

6.6.3.2 Comparative based on DB2

The performances of the existing ensemble classifier and the PSO-optimized ensemble classifier are compared in Fig. 6.5. First, the suggested PSO-optimized ensemble classifier's accuracy improvement rate is

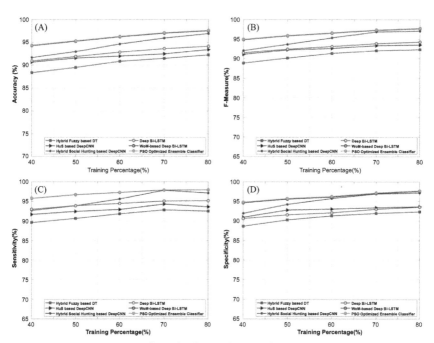

Figure 6.4 Comparative analysis for heart disease dataset (A) accuracy, (B) F-measure, (C) sensitivity, and (D) specificity.

measured and it achieved a rate of 0.08% improvement when compared to the WOM-based deep BiLSTM in Fig. 6.4A. The suggested PSO-optimized ensemble classifier, which achieved a rate of 0.08% improvement when compared to the previous technique, is also measured to determine the improvement rate in terms of F-measure shown in Fig. 6.4B. Similarly, Fig. 6.4C shows the sensitivity is evaluated and an improvement rate of 0.01% is observed. As a result of comparing the final specificity measurement to the previous approach, an improvement rate of 0.13% was observed in Fig. 6.4D.

6.6.4 Comparative discussion

As shown in Tables 6.1 and 6.2, the PSO-optimized ensemble classifier is more effective at detecting the heart disease and improving the classifier performance. As a result, using the Kaggle dataset, the ensemble classifier model outperformed all other models by 97.62%, 97.75%, 97.97%, and 97.66%. The model's achieved efficiency for the dataset for heart disease is 96.71%, 97.33%, 97.99%, and 97.25%, respectively.

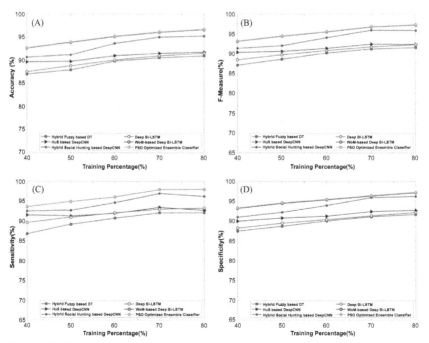

Figure 6.5 Comparative analysis for heart disease dataset (A) accuracy, (B) F-measure, (C) sensitivity, and (D) specificity.

Table 6.1 Comparative performance using the Kaggle dataset.

Methods	DB1			
	TP 80			
	Accuracy	F-Measure	Sensitivity	Specificity
T-1	92.22	92.35	92.51	92.32
T-2	93.43	93.55	93.61	93.62
T-3	96.93	97.05	97.11	97.12
T-4	94.15	94.27	95.15	93.52
T-5	97.53	97.66	97.93	97.52
PSO-optimized ensemble classifier	97.62	97.75	97.97	97.66

Table 6.2 Comparative performance using heart disease dataset.

Methods	DB2			
	TP 80			
	Accuracy	F-Measure	Sensitivity	Specificity
T-1	90.93	91.56	92.09	91.66
T-2	91.75	92.39	92.70	92.71
T-3	95.25	95.89	96.20	96.21
T-4	91.53	92.31	93.25	92.16
T-5	96.63	97.25	97.99	97.13
PSO-optimized ensemble classifier	96.71	97.33	97.99	97.25

6.7 Conclusion

The prediction of heart disease has become an emerging topic among researchers where the influence of technology in medical field is literally important for diagnosing the heart disease. In this research a PSO-optimized ensemble classifier is used to predict heart disease effectively. The modified EDH model is used to securely transfer the IoT environment's collected data via the network in advance. The fusion parameters present in the ensemble classifier model are tuned using the PSO, and the medical data can be transmitted on the network securely using the EDH algorithm with less time. The PSO method accomplished accuracy, *F*-measure, sensitivity, and specificity of 97.62%, 97.75%, 97.97%, and 97.66% for heart disease dataset and for Kaggle dataset 96.71%, 97.33%, 97.99%, and 97.25% are attained. The ensemble classifier can also be used to detect infections early on, allowing for effective treatment to reduce the severity of the condition. In future, more efficient classifiers will be used to increase the prediction accuracy since this PSO-optimized ensemble classifier method prediction is slightly less for real-time applications. This research mainly focuses on heart disease prediction where it does not focus on specified feature selection techniques and in future the model will be designed with advanced feature selection models.

References

[1] M.M. Islam, A. Rahaman, M. Islam, Development of smart healthcare monitoring system in IoT environment, SN Computer Science 1 (2020) 1–11.

[2] H. Pandey, S. Prabha, Smart health monitoring system using IOT and machine learning techniques, in: 2020 sixth international conference on bio signals, images, and instrumentation (ICBSII), IEEE, 2020, pp. 1–4.

[3] M.A. Khan, An IoT framework for heart disease prediction based on MDCNN classifier, IEEE Access 8 (2020) 34717–34727.

[4] S. Basheer, A.S. Alluhaidan, M. AyshaBivi, Real-time monitoring system for early prediction of heart disease using Internet of Things, Soft Computing 25 (18) (2021) 12145–12158.

[5] S.M. Riazul Islam, K.D. Humaun Kabir, M. Hossain, K.-S. Kwak, The Internet of Things for health care: a comprehensive survey, IEEE Access 3 (2015) 678–708. Available from: https://doi.org/10.1109/ACCESS.2015.2437951.

[6] T. Lin, H. Rivano, F. Le Mouel, A survey of smart parking solutions, IEEE Transactions on Intelligent Transportation Systems 18 (2017) 3229–3253. Available from: https://doi.org/10.1109/TITS.2017.2685143.

[7] G. Joyia, R. Liaqat, A. Farooq, S. Rehman, Internet of Medical Things (IOMT): applications, benefits and future challenges in healthcare domain, The Journal of Communication 12 (2017) 240–247. Available from: https://doi.org/10.12720/jcm.12.4.240-247.

[8] M.A. Khan, F. Algarni, A healthcare monitoring system for the diagnosis of heart disease in the IoMT cloud environment using MSSO-ANFIS, IEEE Access 8 (2020) 122259–122269.

[9] F. Ali, S. El-Sappagh, S.M. Riazul Islam, D. Kwak, A. Ali, M. Imran, et al., A smart healthcare monitoring system for heart disease prediction based on ensemble deep learning and feature fusion, Information Fusion 63 (2020) 208–222.

[10] K.V.V. Reddy, I. Elamvazuthi, A.A. Aziz, S. Paramasivam, H.N. Chua, S. Pranavanand, An Efficient prediction system for coronary heart disease risk using selected principal components and hyperparameter optimization, Applied Sciences 13 (1) (2023) 118.

[11] O. Taylan, A.S. Alkabaa, H.S. Alqabbaa, E. Pamukçu, V. Leiva, Early prediction in classification of cardiovascular diseases with machine learning, neuro-fuzzy and statistical methods, Biology 12 (1) (2023) 117.

[12] Kaggle dataset for heart disease, https://www.kaggle.com/sid321axn/heart-statlog-cleveland-hungary-final (accessed on April 2021).

[13] Heart disease dataset, https://archive.ics.uci.edu/mL/datasets/heart + disease (accessed on April 2021).

[14] P. Dhaka, B. Nagpal, WoM-based deep BiLSTM: smart disease prediction model using WoM-based deep BiLSTM classifier, Multimedia Tools and Applications (2023) 1−22.

CHAPTER 7

A thorough analysis on mitigating the risk of gastric cancer using proper nutrition

Danish Jamil[1], Sellappan Palaniappan[1], Sanjoy Kumar Debnath[2], Susama Bagchi[2], Sunil Kumar Chawla[2], Tejinder Kaur[2] and Ankit Bansal[2]

[1]Department of Information Technology, Malaysia University of Science and Technology, Petaling Jaya, Malaysia
[2]Chitkara University Institute of Engineering and Technology, Chitkara University, Punjab, India

7.1 Introduction

Gastric cancer (GC) is a type of cancer that develops in the stomach and is commonly linked to malnutrition and other nutritional deficiencies [1]. As per GLOBOCAN 2020 estimation, it is the sixth most prevalent cancer globally, with over 1 million new cases identified in 2022 and nearly 0.8 million deaths. Adenocarcinoma is the most common type of GC, accounting for up to 85% of all cases [2,3]. GC is a major cause of cancer-related deaths worldwide, and several dietary factors have been implicated in its development. The role of nutrition in reducing the risk of GC has been the subject of extensive research, with a focus on identifying dietary factors that can help prevent the disease. Proper nutrition has been a significant factor in reducing the risk of developing GC, with some dietary factors being identified as risk factors, while others have been found to be a protective shield [4,5].

Malnutrition is a severe problem for GC patients, as it affects the digestion and absorption process and can lead to blockages in the digestive system. It is estimated to occur in up to 60% of GC patients, with the severity rate varying depending on the stage of the tumor [5,6]. Malnutrition can lead to a poor quality of life (QoL), increased risk of infection, longer hospital stays, and financial burden. Dysphagia and GC pain are common symptoms of GC, as well as nausea, vomiting, and intestinal obstruction. Advances in diagnosis and treatment have improved the prognosis for GC patients, and early diagnosis and screening is critical

Securing Next-Generation Connected Healthcare Systems
DOI: https://doi.org/10.1016/B978-0-443-13951-2.00010-6

for improving survival and QoL [7]. Irregular bowel syndrome symptoms like abdominal pain or discomfort, bleeding, hematemesis or melena, dizziness and lightheadedness, weight loss, nausea, vomiting, or bloating are also common in GC patients. GC can be challenging to identify early, as symptoms often do not appear until the disease is in advanced stages. Endoscopy has become a crucial tool in the diagnosis of GC, as it allows for visualization of the GC and identification of abnormal growths. Photofluorography and other scientific techniques can also aid in early detection of GC. Endoscopic guided surgery and new optical technologies such as narrowband imaging have also improved the ability to diagnose GC. Early diagnosis and treatment are important for a better prognosis, and simple and low-cost measures can be used to determine if a person has the disease. Effective prevention measures can also help decrease the incidence of GC [8].

The expected outcome of this research or initiative to increase public awareness of the link between nutrition and GC and to promote healthier dietary choices that can help reduce the risk of developing the disease. Additionally, early detection through screening programs can increase the chances of successful treatment and better patient outcomes. Overall, the expected outcome of our work would be to reduce the incidence of GC, improve patient outcomes, and ultimately decrease the burden of the disease on individuals and society as a whole. Overall, interventions aimed to reduce the risk of GC through proper nutrition should be multifaceted, incorporating education, screening, counseling, policy changes, and community-based programs to promote healthy eating habits and reduce the incidence of the disease [9]. Healthcare practitioners can use patient data to create personalized nutrition plans using modern technologies like machine learning (ML) to reduce the risk of stomach cancer. By gathering and analyzing a patient's health, nutrition, and genetic data, ML algorithms can identify patterns that indicate an increased likelihood of developing stomach cancer and suggest dietary adjustments to reduce that risk. For example, an algorithm can identify if a patient is consuming too much red meat or processed foods and recommend an increase in GC-preventive foods like fruits, vegetables, and whole grains. However, to maintain patient confidence and protect their private information, healthcare providers must establish robust security policies and data protection measures such as upgrading security software, encrypting data in transit and at rest, and limiting access to authorized individuals While there is some research on the use of ML algorithms in healthcare and nutrition, there is a gap in research specifically focused on the integration of nutrition into connected healthcare systems using these algorithms to identify patterns in patient data and provide personalized nutrition recommendations while ensuring the security and privacy of patient data. Additionally, there is a need to evaluate the effectiveness of these algorithms in identifying patterns in patient data that may indicate a higher risk of developing GC [10].

This chapter aims to summarize the available evidence on the association between nutrition and GC risk and highlights the dietary factors that can reduce the risk of developing this deadly disease. The findings of this study have important implications for public health policies aimed at reducing the burden of GC, as well as for individuals who want to adopt a healthy and balanced diet to reduce their risk of developing the disease.

7.2 Research contributions

The chapter aims to:
1. Examining the risk factors associated with the development of GC and how they affect dietary intake, dietary change, and nutritional status in GC patients' QoL.
2. To investigate how risk factors such as dietary intake may contribute to the development of GC and how they may affect the QoL of GC patients.
3. Assess the impact of nutritional status on the QoL of GC patients and, potentially, provide solutions to improve their QoL through a Systematic literature review of the effects of dietary factors in GC patients and timely diagnosis of the incidence of GC.
4. To investigate the potential benefits of integrating proper nutrition into connected healthcare systems using machine learning algorithms.

The chapter is structured as follows: Section 7.2 describes the research methods used for the review. Section 7.3 reviews the literature on the QoL in relation to GC It also describes the material and methods used to explain the methodological procedure. Section 7.4 identifies critical risk factors associated with nutrition and their impact on patient QoL, and contains a detailed discussion of the analysis findings of these risk factors. Section 7.5 discusses the strengths, limitations, and future work of the study. Section 7.6 concludes the chapter.

7.2.1 Methodology

7.2.1.1 Search strategy

This systematic literature review aimed to identify and analyze studies on the association between GC and various GC risk factors. The authors used specific keywords to search the Google Scholar and PubMed databases for relevant articles published between 2015 and 2021. They reviewed and extracted relevant materials using standardized methods and tools, as shown in Table 7.1 where a more detailed description of inclusion and exclusion factors and their explanations is provided. The overall goal of the review was to provide a comprehensive analysis of the current state of knowledge on the topic, which could have implications for prevention and treatment strategies.

We provide the inclusion and exclusion criteria for the systematic literature review. It is important to have clear criteria to ensure that the articles included in the review are relevant to the research question and meet certain standards. By including only articles based on human subjects and published in the English language between 2015 and 2021, the review can focus on recent and relevant studies. The exclusion criteria also help to narrow down the focus of the review and exclude studies that may not meet the standards or are not relevant to the research question. As shown in Table 7.2.

This study selection and data extraction process followed a rigorous and systematic approach. The use of the PRISMA checklist to assess the quality of articles is a standard practice in systematic reviews. The screening of titles and abstracts by one-author and full-text articles by another author reduces the potential for bias in the selection process. Exclusion criteria were clearly defined, and the flow diagram provided a clear

Table 7.1 Criteria (including and excluding) and related justification.

Inclusion	Justification
Published papers in 2015−21 in conference proceedings or journals.	Use the most recent findings only.
Papers present identifying risk factor associated with GC.	Dietary factor associated with GC to reduce the morality rate.
Exclusion	**Justification**
Papers, which are not in English languages. Secondary (review) and tertiary studies are not in focus.	English as an international language is standardized. Concentrate on primary research.

Table 7.2 Search strategy of the research.

Search engines and databases and search date	Search terms
PubMed, Google Scholar, Science Direct, and Elsevier Up to Dec 2022	Strategy: #1 AND #2 AND #3, #1 AND #2 AND #4 AND #5, #1 AND #2 AND #6, #1 AND #2 AND #7, #1 AND #2 AND #8 AND #9, #1 AND #2 AND #10 •#1 GC, #2 Risk factor, #3 Diet, #4 Lifestyle, #5 Environmental factors, #6 Family history, #7 Infections, #8 Treatments, #9 Medical conditions, #10 Demographic characteristics, nutrition

overview of the article selection process. The final selection of 38 articles suggests that the review focused on relevant and high-quality studies related to the research topic as shown in Fig. 7.1.

7.2.1.2 Risk factors and prevention of gastric cancer

Helicobacter pylori is a bacterium that is known to cause chronic gastritis, which can increase the risk of developing GC. However, not all cases of GC are related to *H. pylori* infection. There are various factors that can contribute to the development of GC, including, but not limited to, diet, lifestyle, environmental exposures, family history, medical disorders, infections, demographic traits, and genetic abnormalities. So, while *H. pylori* is a known risk factor for GC, it is not the only cause. The various causes of anti-*H. pylori* GC in the human body are food or nutrition imbalance, lifestyle and environmental exposures, family history, medical disorders, infections, demographic traits, and genetic abnormalities. It emphasizes the importance of early diagnosis and screening as preventative approaches and notes that diets high in fresh fruits and vegetables, low in salt, red and processed meats, and moderate alcohol can reduce GC risk [11]. It also states that smoking and excessive alcohol consumption can increase the risk of GC, as well as exposure to certain environmental factors such as dust, nitrogen oxides, and radiation. Additionally, it notes that certain occupational groups have an increased risk of GC [12]. The passage also mentions that some modifiable factors such as changing dietary habits can have a positive

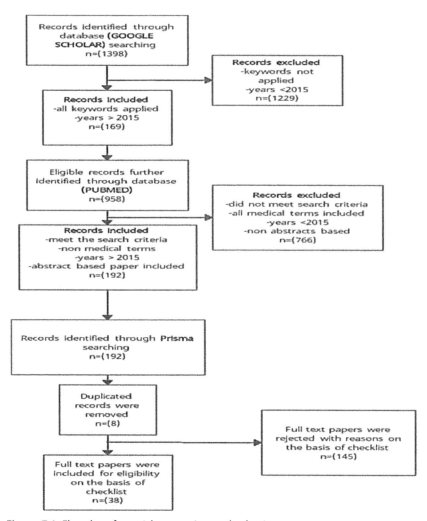

Figure 7.1 Flowchart for article screening and selection.

impact on reducing GC risk and that disparities in GC patients' ethnic and geographic backgrounds, such as immigrants having a greater frequency than native-born citizens, are an important factor to consider in understanding and preventing GC. It concludes that understanding the underlying causes of the disease is critical to the efficacy of preventive initiatives [13].

7.2.1.3 Dietary factors increasing the risk for gastric cancer

Some of the known risk factors associated with GC such as *H. pylori* infection: *H. pylori* is a type of bacteria that can infect the stomach and cause inflammation. Over time, this inflammation can damage the lining of the stomach and increase the risk of developing GC. Family history: A family history of GC or certain inherited genetic mutations can increase the risk of developing the disease.

Age: GC is more common in older adults, with the risk increasing after age 50. Diet: A diet high in salt, processed meat, and low in fruits and vegetables can increase the risk of developing GC. Tobacco use: Smoking or using other tobacco products can increase the risk of developing GC. Alcohol consumption: Heavy alcohol consumption can increase the risk of GC. Obesity: Being overweight or obese can increase the risk of GC. Previous stomach surgery: Individuals who have had certain types of stomach surgery, such as a gastrectomy, may have an increased risk of developing GC. It is important to note that while these factors can increase the risk of GC, they do not necessarily cause GC on their own. Many factors can contribute to the development of GC, and some cases of GC may occur without any identifiable risk factors. Early detection and treatment are important in improving outcomes for those with GC [12,14]. Antioxidant-rich diets, particularly those high in vitamins C and E, have been shown to reduce the risk of GC. However, some studies have yielded inconclusive results. *H. pylori* infection is the most common cause of GC, responsible for 65%−80% of cases, and genetic factors account for around 10% of cases. GC is more common in men than women, and most patients are diagnosed at an advanced stage, resulting in a low 5-year survival rate. Factors such as ethnic and geographic background, dietary habits, and modifiable factors can also play a role in the development of GC. Public health awareness and proper food storage practices have led to a decrease in GC incidence [15].

7.2.1.4 Nutritional status and weight loss

Adequate nutritional care for cancer patients is crucial, as malnutrition can have a significant impact on their health outcomes, QoL, and overall well-being. Malnutrition is a common problem in cancer patients, particularly those with gastrointestinal cancers like GC. Malnutrition can occur due to a variety of reasons, including cancer treatment side effects, tumor-induced anorexia, and reduced nutrient absorption [16]. Appetite-stimulating medications can help cancer patients who suffer from a lack of appetite or cachexia to increase their food intake. In addition, administering parenteral nutrition through infusion can be necessary for patients who are unable to eat or absorb nutrients orally. However, these interventions should be used judiciously and under the guidance of a healthcare professional, as they may have potential side effects or complications. Early and repeated examinations of nutritional status in cancer patients, especially those with gastrointestinal cancers, are essential for identifying malnutrition and developing appropriate intervention strategies. Nutritional support has been shown to improve the QoL and food consumption of cancer patients, and it may even improve treatment outcomes and survival rates. Therefore it is essential to implement a comprehensive approach that involves identifying at-risk patients, monitoring their nutritional status, and providing appropriate interventions to prevent or manage malnutrition. Weight loss and malnutrition are common issues among cancer patients and can negatively affect the QoL. Studies have shown that patients who do not experience weight loss have a better QoL than those who do [17]. Weight loss is often accompanied by symptoms such as fatigue, nausea, vomiting, and lack of appetite. Research has also found that when cancer patients' weight stabilizes, their overall survival rate is higher. Studies have shown that weight loss occurs in 31%– 85% of cancer patients, with those with GC experiencing the highest rates of weight loss [18].

7.2.1.5 Quality of life and nutrition interventions

QoL is a broad term that refers to a person's physical, mental, and social well-being. It is not just a disease-free state but also includes factors such as the ability to cope, have fun, and live independently. Health-related QoL (HR-QoL) is a concept that includes both the physical and psychological effects of an illness and its treatment [7] as shown in Fig. 7.1. Cancer patients often experience symptoms such as fatigue, pain, lack of energy, and loss of appetite, which can affect their HR-QoL. Proper nutrition plays a significant role in improving QoL for cancer patients. Nutritional treatment and counseling should be integrated into overall cancer care to alleviate symptoms and improve response to treatment. There are no universal diets for cancer patients, but basic guidelines such as a diet rich in fruits, vegetables, fats, and oils, with a sugar intake, can be followed. In some cases, gastric tubes or intravenous feeding may be necessary. Nutrition interventions should focus on minimizing symptoms and indicators associated with nutrition diagnoses. Nutrition is a significant factor in the treatment of cancer and depletion is a vital component in the QoL of cancer patients [19].

7.3 Literature review

Maslow's theory of the hierarchy of needs is a model of psychological health that describes how individuals are driven to satisfy their basic needs to move on to higher levels of needs. The five stages of human needs in Maslow's theory are physiological, safety, love and belonging, esteem, and self-actualization. These needs are represented in a pyramid form, with physiological needs at the bottom and self-actualization at the top. A person must address their basic needs before they can move on to the next level of needs. Maslow also proposed a theory of QoL, which states that the greater the amount of need fulfillment a person achieves, the greater their perceived QoL. This theory is still considered conceptual by contemporary psychology specialists [20] as shown Fig. 7.2.

7.3.1 Malnutrition and cancer treatment

It is a significant issue as it affects the effectiveness of treatment and can lead to a reduction in treatment dosage or complete cessation of treatment. Cancer treatments such as surgery, radiation, and chemotherapy alter the body's digestion, absorption, and use of food [22]. Additionally, cancer and its treatment can significantly impact appetite and weight loss, making it challenging for patients to get proper nutrition. Studies have shown that cancer patients are at risk of malnutrition and not getting enough nutritional care. The World Cancer Research Fund (WCRF) Cancer Report recommends consuming a diet rich in fruits and vegetables, legumes, and whole grains to lower the risk of cancer, as well as limiting alcohol consumption and engaging in regular physical exercise [23].

Figure 7.2 Maslow's hierarchy of needs [21].

7.3.2 Consequences of impaired nutritional status

mpaired nutritional status can have various negative effects on both the individual and the healthcare system. Malnutrition can cause changes in body composition, decreased physical and mental performance, and poorer clinical outcomes from illness. It is also associated with an increased risk of disease and mortality, longer hospital stays, decreased tolerance to treatment, and worse survival rates [24]. Malnutrition is a critical prognostic factor in cancer patients and is responsible for 40% of all cancer deaths. In older adults, malnutrition increases the risk of illness, reduces QoL, and increases the likelihood of death. The rising costs of hospitalization and medication can also put additional strain on healthcare systems. Additionally, obesity is also considered a form of malnutrition as it can also cause vitamin deficiencies and increase the risk of chronic disease and mental health and social issues. Dietary treatments for disease prevention and management require collaboration between patients and doctors, and factors such as treatment intensity, duration, and individual risk factors must be taken into account. Cancer patients may experience weight loss before diagnosis and may face long-term health issues after surgery or radiation therapy. Chemotherapy can also lead to dietary issues, and the likelihood of these issues varies among patient groups [25].

7.3.3 Factors affecting diet and nutritional status

There are several factors that can impact appetite and nutrition in older adults, including changes in physiology, food safety and hygiene practices, health behaviors, medication, and psychological factors. Age-related changes such as delayed GC emptying can lead to reduced appetite and energy intake, which can lead to poor diet quality and diversity [26]. Older adults are also more susceptible to foodborne illnesses and infections due to compromised immune systems and nutritional deficiencies. Health behaviors such as smoking, inactivity, excessive alcohol use, and poor diet also congregate among the older population and contribute to noncommunicable diseases. Medications can also impact nutritional status, either directly or indirectly through effects on appetite. Psychological factors such as depression can also play a role in the nutritional status of older adults [27].

7.4 Discussion and findings

It is important to understand the risk factors associated with the development of GC to identify individuals at high risk and implement preventive measures. A systematic review was conducted to identify potential risk factors for GC, which were grouped into six categories: diet, lifestyle, genetic predisposition, family history, medical disorders, infections, and demographic variables.

1. Diet and lifestyle-related factors, such as high red and processed meat consumption, alcohol consumption, and tobacco use.
2. Genetic and familial factors, such as inherited genetic mutations and a family history of GC. Medical conditions and infections, such as inflammatory bowel disease, *H. pylori* infection, and type 2 diabetes.
3. Demographic factors, such as age, sex, and race. Environmental factors, such as exposure to pollutants and pesticides.
4. Other factors, such as obesity, sedentary lifestyle, and low intake of fruits and vegetables.

Table 7.3 provides specific examples of each of these risk factors and the level of evidence supporting their association with GC Dietary factors such as high salt intake, low fruit and vegetable consumption, and the consumption of processed and red meat were identified as potential risk factors for GC. Lifestyle factors such as smoking and alcohol consumption also found to be associated with an increased risk of GC. Genetic predisposition, including specific gene mutations and family history of GC, were identified as potential risk factors. Medical disorders such as chronic atrophic gastritis and *H. pylori* infection identified as potential risk factors for GC. Demographic variables such as male gender, older age, and ethnicity were also found to be associated with an increased risk of GC [28,29].

The systematic review identified a range of potential risk factors for GC, which used to identify individuals at high risk and implement preventive measures. These measures may include lifestyle modifications, such as smoking cessation and reducing alcohol consumption, and dietary changes, such as increasing fruit and vegetable consumption and reducing the consumption of processed and red meat. In addition, screening for *H. pylori* infection and other medical conditions associated with an increased risk of GC may be beneficial in identifying individuals at high risk.

7.4.1 Diet

The diet and dietary habits of an individual can play a significant role in the development of GC. Excessive salt consumption and a poor diet with a lack of nutrition have been linked to an increased risk of GC. Eating

Table 7.3 Critical risk factor for gastric cancer.

S. no.	The risk factor	Subsections	Description of the product	Factors with high risk/factors with low risk
1	Nutritional or dietary	Consumption of salt and diets high in salt	Preserved vegetables and other items.	Risk is an increase
		Spicy cuisines		
		Meat (the red smoked processed salty) foods containing milk	Fried meals	Risk is an increase
			Unusual eating habits	Risk is an increase
		Fermentation of salt-cured and smoked fish Tea is served hot.	Foods high in starch and sugar	Risk is an increase
			Hot meals	Risk is an increase
		Leftover and moldy bread and other stuff Consuming a low-vitamin C diet (vitamin C deficiency)	Lack of access to drinking water	Risk is a decrease
			N-nitroso	Risk is an decrease
			Fatty and oily substances	Risk is an increase
		Fresh fruits and vegetables are not being consumed in sufficient quantities.	Frozen foods that have been steamed, boiled foods that are high in fat	Risk is an increase
			Granular product that has been refined	Risk is an increase
2	Factors such as lifestyle and the surrounding environment	Smoking	Opium anxiety	Risk is an increase
		Lack of physical exercise and inactivity	Glass pipe	Risk is an increase
			Get daily surplus energy	Risk is an increase
		Overeating and eating too quickly	Cement	Risk is an increase
		Exposure to work-related hazards	Mineral dust	Risk is an increase
			Radiation-induced	Risk is an increase
		An ionizing radiation source	Chromatid VI	Risk is an increase

fast, cooking with oil, eating hot or spicy food, and eating high-glycemic-index foods have all been associated with a higher risk of GC. Consuming more fruits and vegetables, on the other hand, has been shown to lower the risk of GC. Vegetarians have also been found to have a decreased chance of developing GC compared to meat-eaters [30]. Additionally, vitamin C has been shown to have protective effects against GC by suppressing the synthesis of N-nitroso compounds and reducing the likelihood of precancerous lesions. Drinking water sources can also be a risk factor for GC, as contamination with bacteria, cyanotoxins, sulfates, nitrates, and other contaminants can lead to intestinal cancer. Hot liquids like tea and coffee have also been linked to an increased risk of GC. It is suggested that individuals should limit their intake of hot liquids and increase their consumption of fruits and vegetables to lower their risk of GC. The studies discussed in this passage indicate that several factors may increase a person's risk of developing GC. These include a family history of the disease, infection with the bacteria *H. pylori*, infection with non-*H. pylori*-helicobacters, infection with the Epstein–Barr virus, certain medical conditions and treatments (such as gastrectomy and GC surgery), blood type (A), menopause and older age, and a history of gastric ulcers. The passage also states that these risks may vary depending on the specific study and population being examined, and that more research is needed to fully understand the molecular genetic pathways of GC [15]. Healthcare professionals can identify and address the nutritional needs of GC patients through nutritional assessment, counseling, multidisciplinary care, addressing treatment-related side effects, nutritional supplementation, and monitoring and follow-up. These interventions can improve treatment outcomes and QoL for patients with GC.

7.5 Recommendations for practice future development and challenges

Nutrition plays a critical role in cancer treatment and accurate identification of a patient's nutritional status is important for devising an appropriate nutritional intervention. Clinical dietitians who specialize in nutritional oncology can provide high-quality nutrition advice and guidance to cancer patients. Studies have shown that patients who receive nutritional counseling throughout chemotherapy are less likely to lose significant weight and have a higher overall QoL [31]. The primary goals of nutritional intervention in cancer treatment are to prevent early mortality, reduce complications, and enhance life expectancy. Oral nutritional supplements are often preferred over parenteral nutrition due to their lower cost and ease of administration. Emotional support for cancer patients is also important during this challenging time. Studies have shown that the cancer stage is the most critical factor in determining a patient's QoL and those nutritional treatments should begin as soon as cancer is identified. Overall, reducing the risk of GC through proper nutrition has significant social impacts, including improved public health, increased awareness of the importance of nutrition, reduction of healthcare costs, improved QoL, and environmental benefits [32].

7.5.1 Recommendations for research

The American Society of Clinical Oncology recommends that patients with GC speak with oncology-trained dietitians while undergoing treatment. Supportive and appropriate dietary methods are essential for medical personnel caring for cancer patients receiving chemotherapy. More research is needed to determine the best meals for GC patients to reduce weight loss. Nutritional treatment for GC patients requires medical professionals to understand their roles and duties and to educate patients about their condition and available treatment options. Surgery, radiation, and chemotherapy are options available to those suffering from GC with the hope of curing them, extending their lives, and alleviating their symptoms [33]. However, these treatments can have a detrimental impact on HR-QoL and can lead to a greater risk of developing mental comorbidities. The study has a thorough evaluation of databases over a considerable period but does not include animal or cellular investigations, which are important in understanding the fundamentals of cancer research and associated theories. Our objectives focus on understanding the most effective strategies for healthcare professionals to identify and address the nutritional needs of GC patients to improve their treatment outcomes and QoL. By achieving these objectives, we can develop evidence-based recommendations for healthcare professionals and policy-makers to improve the nutritional care of patients with GC. In addition, interconnected healthcare systems can leverage ML and other cutting-edge technologies to better serve patients. For instance, virtual reality apps can aid in pain treatment and rehabilitation, telehealth systems can enable remote consultations and monitoring, and wearable devices can gather real-time biometric data [34].

ML algorithms can also help predict security breaches by analyzing patterns in patient data and alerting healthcare providers to potential threats. Healthcare professionals can use these predictive models to detect phishing and malware attacks based on user behavior patterns, taking preventative measures to protect patients and their data. To maintain the safety and privacy of interconnected healthcare systems, strong security standards, and data protection measures are crucial. This includes restricting access to sensitive data, implementing encryption during processing, and anonymizing data to avoid patient identification. By incorporating ML algorithms into interconnected healthcare systems to promote healthy eating and reliable safety measures, healthcare professionals can improve patient outcomes and reduce the risk of stomach cancer [35].

7.5.2 Strength and limitations of study

This research on the relationship between nutrition and GC treatment has several limitations, including the fact that it is based on a limited number of studies and the quality of the papers included in the review is highly dependent on the quality of the systematic review itself. Additionally, the equipment used in each study was unique, making it difficult to draw meaningful comparisons between them. The research also has to overcome "publication bias" which is the inability to obtain published or under consideration for publication before it was complete, which is a problem shared by all systematic review studies. However, the greatest strength of this research is its thorough database scanning and identification of multiple influencing variables. The research is also limited by the fact that it only includes studies published in English, which might not capture all relevant research conducted in other languages [36]. Despite these limitations, the findings of this research can provide valuable insights for future work on the relationship between nutrition and GC treatment. In summary, this research highlights the importance of assessing the impact of treatment on HR-QoL in patients with GC, as it is a severe public health concern with a poor prognosis. The systematic review had several limitations, including the inability to access all studies due to publication bias and the diversity of research methods and equipment used. However, the review found a strong correlation between nutritional status and QoL in cancer patients and identified opportunities for future research in areas such as determining best management practices for nutritional assessment and intervention and promoting a holistic and patient-centered approach to cancer treatment [37,38].

7.6 Conclusion

GC remains a prevalent disease with a poor prognosis, largely linked to *H. pylori* gastritis. However, promoting a healthy lifestyle with adequate nutrition, moderation in alcohol consumption, and avoiding smoking can help lower the incidence of GC. Eating fruits and vegetables is associated with a lower incidence of GC, but it is unclear if all fruits and vegetables are equally effective. Micronutrients such as selenium, iron, zinc, vitamin C, and folate have also been investigated. A high-fat diet increases the likelihood of developing GC, but fresh fruits and vegetables and certain micronutrients are protective. GC risk may also be influenced by factors such as drinking hot water, consuming pickled foods, and exposure to *H. pylori*. Decreasing GC-related mortality requires early detection and treatment. The research aimed to evaluate the effect of dietary determinants on the QoL of GC patients and recommends nutritional counseling at the time of diagnosis to improve patients' QoL.

Additionally, early detection and treatment are crucial to decrease GC-related mortality. The integration of ML algorithms into healthcare systems can identify dietary patterns that increase the risk of stomach cancer and suggest dietary adjustments to prevent its development. Nutritional counseling at the time of diagnosis can also improve the QoL of GC patients. It is important for healthcare providers to prioritize patient privacy and data protection while utilizing modern technologies to improve patient outcomes and reduce the risk of GC.

References

[1] P. Rawla, A. Barsouk, Epidemiology of gastric cancer: global trends, risk factors and prevention, Przeglad Gastroenterologiczny 14 (1) (2019) 26.
[2] B.S. Chhikara, K. Parang, Global Cancer Statistics 2022: the trends projection analysis, Chemical Biology Letters 10 (1) (2023) 451.
[3] F. Taleghani, M. Ehsani, S. Farzi, S. Farzi, P. Adibi, A. Moladoost, et al., Nutritional challenges of gastric cancer patients from the perspectives of patients, family caregivers, and health professionals: a qualitative study, Supportive Care in Cancer: Official Journal of the Multinational Association of Supportive Care in Cancer 29 (7) (2021) 3943−3950.
[4] R. Sitarz, M. Skierucha, J. Mielko, G.J.A. Offerhaus, R. Maciejewski, W.P. Polkowski, Gastric cancer: epidemiology, prevention, classification, and treatment, Cancer Management and Research 10 (2018) 239.
[5] M. Planas, J. Álvarez-Hernández, M. León-Sanz, S. Celaya-Pérez, K. Araujo, De, et al., Prevalence of hospital malnutrition in cancer patients: a sub-analysis of the PREDyCES®study, Supportive Care in Cancer: Official Journal of the Multinational Association of Supportive Care in Cancer 24 (1) (2016) 429−435.
[6] Z. WenRong, W. Li, et al., Effect of enteral and parenteral nutritional diet intervention on recovery, nutritional status and immune function after radical operation of gastric cancer, Journal of Hainan Medical University 25 (1) (2019) 58−64.
[7] K. Schütte, C. Schulz, K. Middelberg-Bisping, Impact of gastric cancer treatment on quality of life of patients, Best Practice & Research Clinical Gastroenterology 50−51 (Suppl 6) (2021) 101727.

[8] M.R.A. Fujiyoshi, H. Inoue, Y. Fujiyoshi, Y. Nishikawa, A. Toshimori, Y. Shimamura, et al., Endoscopic classifications of early gastric cancer: a literature review, Cancers (Basel) 14 (1) (2021) 100.

[9] C. Yu, E.J. Helwig, The role of AI technology in prediction, diagnosis and treatment of colorectal cancer, Artificial Intelligence Review 55 (1) (2022) 323–343.

[10] S. Tripathy, D.K. Verma, M. Thakur, N. Chakravorty, S. Singh, P.P. Srivastav, Recent trends in extraction, identification and quantification methods of *Centella asiatica* phytochemicals with potential applications in food industry and therapeutic relevance: a review, Food Bioscience 49 (2022) 101864.

[11] T. Zhang, H. Chen, X. Yin, Q. He, J. Man, X. Yang, et al., Changing trends of disease burden of gastric cancer in China from 1990 to 2019 and its predictions: findings from Global Burden of Disease Study, Chinese The Journal of Cancer Research 33 (1) (2021) 11.

[12] A. Shahi, V.P.B. Koyyala, E.S. Rathaur, M.A. Biddut, A. Hossain, M.K. Hasan, et al., Association between gastric cancer with behavioral and dietary factors: a hospital based case-control study in South Asia, Asian Journal of Oncology (2021).

[13] J.A.E. Langius, M.C. Zandbergen, S.E.J. Eerenstein, M.W. van Tulder, C.R. Leemans, M.H.H. Kramer, et al., Effect of nutritional interventions on nutritional status, quality of life and mortality in patients with head and neck cancer receiving (chemo) radiotherapy: a systematic review, Clinical Nutrition (Edinburgh, Scotland) 32 (5) (2013) 671–678.

[14] S.K. Shah, D.R. Sunuwar, N.K. Chaudhary, P. Rai, P.M.S. Pradhan, N. Subedi, et al., Dietary risk factors associated with development of gastric cancer in Nepal: a hospital-based case-control study, Gastroenterology Research and Practice 2020 (2020).

[15] M.C. Tan, N. Mallepally, Q. Ho, Y. Liu, H.B. El-Serag, A.P. Thrift, Dietary factors and gastric intestinal metaplasia risk among US veterans, Digestive Diseases and Sciences 66 (5) (2021) 1600–1610.

[16] S. Ida, K. Kumagai, S. Nunobe, Current status of perioperative nutritional intervention and exercise in gastric cancer surgery: a review, Annals of Gastroenterological Surgery 6 (2) (2022) 197–203.

[17] D.W. Pekmezi, T.E. Crane, R.A. Oster, L.Q. Rogers, T. Hoenemeyer, D. Farrell, et al., Rationale and methods for a randomized controlled trial of a dyadic, web-based, weight loss intervention among cancer survivors and partners: the DUET Study, Nutrients 13 (10) (2021) 3472.

[18] S.C. Regueme, I. Echeverria, N. Monéger, J. Durrieu, M. Becerro-Hallard, S. Duc, et al., Protein intake, weight loss, dietary intervention, and worsening of quality of life in older patients during chemotherapy for cancer, Supportive Care in Cancer: Official Journal of the Multinational Association of Supportive Care in Cancer 29 (2021) 687–696.

[19] S.K. Rupp, A. Stengel, Influencing factors and effects of treatment on quality of life in patients with gastric cancer—a systematic review, Front Psychiatry 12 (2021). Available from: https://www.frontiersin.org/articles/10.3389/fpsyt.2021.656929/full.

[20] T. Bridgman, S. Cummings, J. Ballard, Who built Maslow's pyramid? A history of the creation of management studies' most famous symbol and its implications for management education, Academy of Management Learning & Education 18 (1) (2019) 81–98.

[21] C.-Y. Shih, C.-Y. Huang, M.-L. Huang, C.-M. Chen, C.-C. Lin, F.-I. Tang, The association of sociodemographic factors and needs of haemodialysis patients according to Maslow's hierarchy of needs, Journal of Clinical Nursing 28 (1–2) (2019) 270–278.

[22] L.-B. Xu, M.-M. Shi, Z.-X. Huang, W.-T. Zhang, H.-H. Zhang, X. Shen, et al., Impact of malnutrition diagnosed using Global Leadership Initiative on Malnutrition

criteria on clinical outcomes of patients with gastric cancer, Journal of Parenteral and Enteral Nutrition 46 (2022) 385–394.

[23] I.M. de Sousa, F.M. Silva, A.L.M. de Carvalho, I.M.G. da Rocha, A.P.T. Fayh, Accuracy of isolated nutrition indicators in diagnosing malnutrition and their prognostic value to predict death in patients with gastric and colorectal cancer: a prospective study, Journal of Parenteral and Enteral Nutrition 46 (2021) 508–516.

[24] C.A. Santos, I.M. Santos, L. Mendes, H. Mansinho, Gastric cancer: nutritional and functional status & survival time/mortality, Clinical Nutrition ESPEN 46 (2021) S720–S721.

[25] S. Garrosa Muñoz, J. López Sánchez, O. Abdel-Lah Fernández, I. Hernández Cosido, I. Jiménez Vaquero, S. Carrero García, et al., Influence of nutritional status on the survival of patients undergoing gastric cancer surgery, The British Journal of Surgery 108 (Supplement 3) (2021) znab160.033.

[26] S.E. Oh, M.-G. Choi, J.-M. Seo, J.Y. An, J.H. Lee, T.S. Sohn, et al., Prognostic significance of perioperative nutritional parameters in patients with gastric cancer, Clinical Nutrition (Edinburgh, Scotland) 38 (2) (2019) 870–876.

[27] Y. Qian, H. Liu, J. Pan, W. Yu, J. Lv, J. Yan, et al., Preoperative controlling nutritional status (CONUT) score predicts short-term outcomes of patients with gastric cancer after laparoscopy-assisted radical gastrectomy, World Journal of Surgical Oncology 19 (1) (2021) 1–10.

[28] M.C. Tan, Q. Ho, T.H. Nguyen, Y. Liu, H.B. El-Serag, A.P. Thrift, Risk score using demographic and clinical risk factors predicts gastric intestinal metaplasia risk in a US population, Digestive Diseases and Sciences 67 (9) (2022) 4500–4508.

[29] S.A. Mahmoodi, K. Mirzaie, M.S. Mahmoodi, S.M. Mahmoudi, A medical decision support system to assess risk factors for gastric cancer based on fuzzy cognitive map, Computational and Mathematical Methods in Medicine 2020 (2020).

[30] J. Kim, A. Oh, H. Truong, M. Laszkowska, M.C. Camargo, J. Abrams, et al., Low sodium diet for gastric cancer prevention in the United States: results of a Markov model, Cancer Medicine 10 (2) (2021) 684–692.

[31] M.C. Mentella, F. Scaldaferri, C. Ricci, A. Gasbarrini, G.A.D. Miggiano, Cancer and Mediterranean diet: a review, Nutrients 11 (9) (2019) 2059.

[32] G. Padalkar, R. Mandlik, S. Sudhakaran, S. Vats, S. Kumawat, V. Kumar, et al., Necessity and challenges for exploration of nutritional potential of staple-food grade soybean, Journal of Food Composition and Analysis (2022) 105093.

[33] C. Rodriguez-Garcia, C. Sánchez-Quesada, E. Toledo, M. Delgado-Rodriguez, J.J. Gaforio, Naturally lignan-rich foods: a dietary tool for health promotion? Molecules (Basel, Switzerland) 24 (5) (2019) 917.

[34] Y. Xu, X. Liu, X. Cao, C. Huang, E. Liu, S. Qian, et al., Artificial intelligence: a powerful paradigm for scientific research, Innov. 2 (4) (2021) 100179.

[35] T.-H. Kim, I.-H. Kim, S.J. Kang, M. Choi, B.-H. Kim, B.W. Eom, et al., Korean practice guidelines for gastric cancer 2022: an evidence-based, multidisciplinary approach, Journal of Gastric Cancer 23 (1) (2023) 3–106.

[36] D. Jamil, S. Palaniappan, S.S. Zia, A. Lokman, M. Naseem, Reducing the risk of gastric cancer through proper nutrition—a meta-analysis, International Journal of Online and Biomedical Engineering 18 (7) (2022).

[37] D. Jamil, Diagnosis of gastric cancer using machine learning techniques in healthcare sector: a survey, Informatica 45 (7) (2022) 147–166.

[38] S. Bagchi, K.G. Tay, A. Huong, S.K. Debnath, Image processing and machine learning techniques used in computer-aided detection system for mammogram screening—a review, International Journal of Electrical and Computer Engineering (IJECE) 10 (3) (2020) 2336–2348.

CHAPTER 8

Application of blockchain and fog computing in healthcare services

Greeshmitha Vavilapalli, Vikash Kumar and Sushruta Mishra
Kalinga Institute of Industrial Technology, Deemed to be University, Bhubaneswar, Odisha, India

8.1 Introduction

In recent years, the world has experienced urbanization, which has resulted in the rising world population. Currently, 54% of people live in cities and 46% in rural areas, and this proportion is expected to rise to 66% by 2050. Modern technologies play a crucial role in this trend, especially new technologies that reduce costs by optimizing the use of resources and prioritize environmentally friendly design [1]. In this regard, Karale and Ranaware [2] and many other researchers have said that big data, Internet of Things (IoT), artificial intelligence, machine learning, multimedia, blockchain, cyber-physical systems, and cloud are the areas of interest [3−8]. The application of this technology is expected to help develop a smart and safe city [9−14]. Blockchain is a decentralized and dispensed ledger technology that securely records transactions across multiple nodes in a network. It makes use of cryptography to ensure statistics integrity, immutability, and transparency. Blockchain has capability packages in numerous industries, such as finance, supply chain management, healthcare, and more.

Fog computing (FC) is a computing paradigm that extends cloud computing to the threshold of the network, toward the statistics source. Its objective is to triumph over the demanding situations of latency, bandwidth usage, and data privacy by processing and reading data locally at the threshold gadgets or gateways. FC permits real-time data processing, faster response times, and advanced capability, making it perfect for packages inclusive of IoT, clever cities, self-sustaining vehicles, and business automation. It provides a decentralized computing method that enhances conventional cloud computing, taking into consideration green and powerful data management and evaluation at the threshold of the network.

Securing Next-Generation Connected Healthcare Systems
DOI: https://doi.org/10.1016/B978-0-443-13951-2.00003-9

Blockchain and FC are rising technologies that have the potential to revolutionize healthcare services. Blockchain, as a decentralized and disbursed ledger, can enhance records security, privacy, and interoperability in healthcare with the aid of developing a tamper-proof and obvious file of patient records, scientific data, and transactions. FC, on the other hand, can allow real-time information processing, analysis, and selection-making at the threshold of the community, closer to patients and healthcare providers. The integration of blockchain and FC in healthcare services can yield several benefits. For instance, patient data can be securely saved and shared among various stakeholders, including hospitals, clinics, pharmacies, and insurers, with the use of blockchain technology. Simultaneously, FC can permit fast data processing and analysis at the threshold gadgets or gateways, allowing for timely diagnosis, treatment, and monitoring of patients. Additionally, the decentralized nature of blockchain and FC enhances data privacy, reduces the risk of data breaches, and ensures data integrity.

According to Treiblmaier et al. [15], a "smart city" is a densely populated area that uses emerging data and communication technologies to connect the physical components, which improves the flow of information, enhances the overall efficiency of the city operations, and elevates the quality of life for the citizens. The fundamental concept is to develop a setting where technology is completely integrated into the community to address urbanization-related issues, particularly by introducing expensive services. According to Burnes and Towers [16], the complexity of the technical and infrastructural systems is what makes smart cities unique. Al-Azzam and Alazzam [17] contend that smart cities are made up of a variety of elements, one of which is intelligent healthcare, which has earned significant recognition in recent years, notably for ensuring the provision of excellent medical care to the populace [18]. The sharing of scientific data on patients between insurance companies and healthcare providers has significantly increased in the healthcare sector, which has led to the rise of healthcare models that are driven by statistics [19]. Strong security measures, such as access limits, are necessary to protect people's privacy because healthcare services generate a significant quantity of statistics [20]. This has increased the need for developing technologies such as the IoT, blockchain, and FC to ensure faster and secure access to healthcare information, thereby improving the quality of healthcare offerings [21]. The current study aims to look at the impact of blockchain and FC, specifically in the framework of healthcare facilities in smart cities. Such initiatives are

likely to alter the established structure of smart cities, giving the locals better access to healthcare services. However, Dwivedi et al. [21] have
pointed out that hackers frequently try to get access to medical records for
identity theft because clinical information is often found in unusual documents such as reviews, videos, images, and raw data, whose integrity is
vital. The blockchain is inexpensive, despite the fact that its security relies
on the evidence of work principle, according to which a transaction is
only deemed genuine if enough computational effort has been made to
fix the cryptographic puzzles, which might be achieved via the authorizing nodes. The incapacity to delete or change the information present in
a block has subsequently impacted blockchain because of the modern
generation which may be exploited for the healthcare system in smart cities [22−24]. Blockchain has been described by Zheng et al. [25] as the
distributed database of transactions on an extensible network that the
authorized parties may only access online. The judicial proceedings confirm the transactions kept on online public databases. Additionally, the
transaction data that are kept on the blockchain are protected by encryption and, as a result, cannot be deleted in any manner. It is hence well
protected against loss and change. Each transaction block is linked to
every other block in a chain, creating a sequence of blocks. It implies that
the likelihood of statistics being changed lowers as the variety of transaction blocks broadens [26]. The transaction information is converted by
applying the hashing in an encrypted form. In case of any trade made to
the previous block, the hash of the block modifies the hash of the subsequent blocks. This is why any exchange made to any block can easily be
noticed and rejected. The procedure of the chained block is proven in
Fig. 8.1.

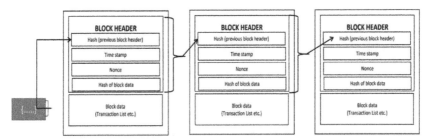

Figure 8.1 Blocks in blockchain. *From D. Yaga, P. Mell, N. Roby, K. Scarfone, Blockchain Technology Overview, NIST, Gaithersburg, MD, 2018.*

The IoT is described by Atzori et al. [27] as a network of gadgets that communicate with one another through M2M communications, enabling the gathering and sharing of data. The technology incorporates a variety of sensors, the cloud, products and services, systems, and communication and network equipment [28]. These technologies make extensive data sharing possible by utilizing their computer power and connectivity [29]. Different layers of technology and communication tools make up the IoT architecture. The IoT system does not have a set architecture that is widely accepted. IoT systems can differ from one another depending on the variety of sensors, communication tools, networks, physical objects, etc. [30].

8.2 Role of blockchain and fog computing in healthcare

FC has been touted as the promising technology for low-cost remote monitoring, cutting down on latency and boosting efficiency, while blockchain has been touted as the promising technology for guaranteeing the protection of private data, building a decentralized database, and improving data interoperability.

FC improves mobility, security, privacy, minimal latency, and network bandwidth by distributing storage, processing, and communications to edge devices, which puts cloud features closer to end users. Applications that are latency-sensitive or real-time are best suited for FC. The architecture of healthcare system based on FC is shown in Fig. 8.2.

The current study focuses on analyzing the impact of blockchain and cloud computing, especially in the context of smart city health. This particular issue is important because the urban economy needs the integration of smart technology to improve health. The cloud computing model is shown in Fig. 8.3. This project should lead to the creation of smart cities that will provide better healthcare to citizens. However, hackers often try to obtain health information for theft, saying that the integrity of medical information is paramount, as it is available in a variety of formats such as reports, videos, photos, and raw data. Since the emergence of Bitcoin, the possibilities to leverage blockchain have become endless as the technology can provide many benefits.

Blockchain is low-cost and its security is based on proof-of-work, which is considered valid only when enough computational work has been done to solve the cryptographic puzzle completed by the permission nodes. The inability to delete or change the information contained in

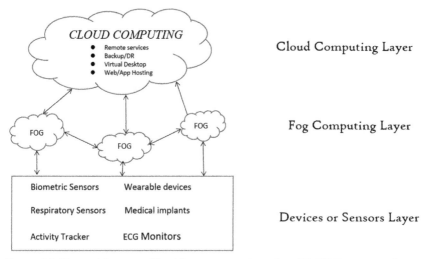

Figure 8.2 The architecture of healthcare system based on FC. *FC,* Fog computing.

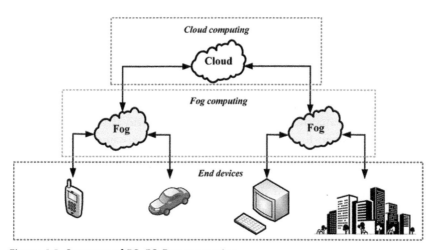

Figure 8.3 Structure of FC. *FC,* Fog computing.

blocks ultimately makes blockchain the best available technology for smart city healthcare. FC can provide rapid data collection and analysis at the edge or gateway, enabling timely diagnosis, treatment, and patient care. In addition, the deployment of blockchain and cloud computing can improve data privacy, reduce the risk of data breaches, and ensure data integrity. Smart cities are densely populated areas that use new information and communication technologies to connect physical devices, thereby improving the flow of information, improving all jobs in the city, and improving the quality of life of citizens. The main idea is to create an environment where technology is successfully applied in the city to cater to the city's economy, especially by adding high-quality services.

8.3 The proposed approach for healthcare security

Consider a situation where multiple CCTV cameras are used to monitor the streets or patients in a hospital room. The video streams captured by the cameras are first sent to local servers that work to filter out places where there may be some significant human activity. The stream subset filter in the first stage is transmitted over the local area network from the local server to the cloud server deployed at the edge of the network. The cloud server can do more to analyze human activities and immediately confirm if something is important. If further computing is required and cannot be computed on the cloud server, only the corresponding video stream is uploaded to the remote cloud computing database on the Internet. FC not only reduces the latency of uploading large video streams to cloud servers but also the bandwidth consumption of the Internet and the associated financial costs. Fig. 8.4 presents the proposed blockchain-based cloud computing healthcare framework for knowledge-based activities. Identifying an action and then recognizing the action as a particular group are two important tasks that must be performed by any cognitive function. The main features are the same, where the same order is issued from each employee for the order to be confirmed. Likewise, features that are unique and can also be calculated effectively in real time are extracted to be sure and efficient. In the proposed method, feature descriptors are obtained from the keyframes of the training video dataset. To minimize the comparison of classification of new features from existing ones, descriptive items are divided into functional class groups or groupings representing important items. In classification, various types of support vector machines (SVMs) are used as vector content of main features to determine the group function of the frame in the video. Based on the quality of service (QoS) requirements of the end-user application, the availability of cloud computing near the base of video production, and the access to the cloud server on the Internet, these actions are geared toward identifying human activities, which must be done on a cloud server.

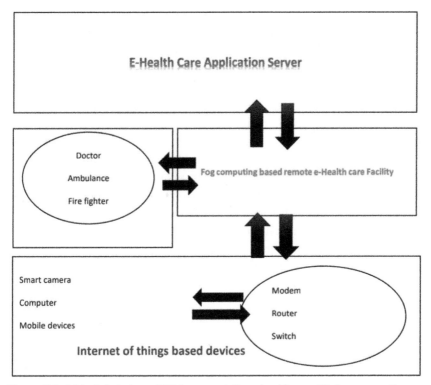

Figure 8.4 A blockchain-based FC framework for e-healthcare. *FC,* Fog computing.

8.3.1 Feature detection and extraction

Calculating the correlation between identical scenes that are shown in two separate visual frames is a crucial step in the action recognition process. The identification of important locations in the picture frames is done using the speed-up robust characteristics (SURF) in the suggested method [31]. The translation, rotation, size, lighting, and contrast of an image have no effect on SURF descriptors. It offers a distinctive description of visual characteristics. Through the use of integral image generation and Hessian matrix approximation, each key point is locally calculated with respect to its neighborhood.

The integral image $S\sum(x, y)$ is determined by summing all pixels in a small rectangular region in the image I, that is,

$$S\sum(x, y) = \sum_{i=0}^{i \le y} \sum_{j=0}^{j \le y} I(i, j) \tag{8.1}$$

The points of interest are located in integral images using the Hessian matrix. At any given point $t = (x, y)$ in the integral image $S\sum(t)$, the Hessian matrix $H(t, \sigma)$ at scale σ is calculated by:

$$H(t, \sigma) = \begin{bmatrix} Lxx(t, \sigma) & Lxy(t, \sigma) \\ Lxy(t, \sigma) & Lyy(t, \sigma) \end{bmatrix} \tag{8.2}$$

Here $L(t, \sigma)$ denotes the convolution of the integral image S's second-order derivatives following its smoothing with a Gaussian filter using the smoothing factor. The determinant of the Hessian matrix is used to calculate the changes in pixels, and only those pixels that provide the highest determinant are selected. SURF uses box filters of various sizes to find locations in the picture that are scaled differently. Haar wavelet responses in the horizontal and vertical directions are computed for a neighborhood of size 6 seconds (where s is the scale at which the point of interest is picked) surrounding the point of interest, providing a descriptor represented by a size n vector.

8.3.2 Action vocabulary construction

The feature vector representing a specific action is utilized for classification abstracted into the action vocabulary. In other words, novel attribute vectors from fresh movies are connected to the feature vectors derived from the clips in the training activity datasets. A simple comparison with a new query descriptor is almost unfeasible in real time because there are so many descriptors for each action in training courses. Partitioning the descriptors into appropriate groups based on the similarities would be a very helpful option. The feature retrieved using SURF from every frame of the films from training class activities yields a relatively big collection in the suggested manner. To increase the categorization process' effectiveness and precision, by applying the k-means clustering approach, the feature descriptions are divided and quantized [32]. In this method, features ($x1$, $x2, \ldots, xn$), acquired as a consequence of SURF, are iteratively partitioned into $k(n)$ clusters ($\acute{a} = c1, c2, \ldots, ck$). Each of the features is of d dimensions. The primary basis for this partitioning is the separation of every feature of the cluster unit. A specific action is represented by the center cc of each cluster (ci), which is referred to as a visual language or action vocabulary.

As seeds or original cluster centers, k random sites are chosen in this situation in the beginning. Then, each feature point is iteratively assigned to the closest cluster center based on the smallest geometric distance determined by applying the Euclidean measurement provided as:

$$d = \sqrt{\sum_{j=1}^{k} \sum_{i=0}^{n} (xi - cj)2} \tag{8.3}$$

Finally, an adjustment to the cluster center takes place by taking the mean of all points in each cluster. The use of k-means is motivated by the fact that k-mean algorithm partitions the features in a finite number of iterations.

8.3.3 Classification

The action vocabulary is constructed by creating clusters that are based on the assignment of feature descriptors representing actions. The centers of these clusters represent a feature descriptor, which is a representation of a particular action. Thus, the problem of action recognition is transformed into a multiclass supervised learning problem that may be solved by any classification algorithm [33].

There are a lot of classification algorithms in the literature, but in the proposed approach for the classification of actions, the SVM-based algorithm is used. The reasons behind the selection of SVM for the action classification are evident from the fact that SVM is more simple and efficient, and gives high accuracy in terms of recognition. SVM is aimed at finding the hyperplane that separates the binary classes with maximum margins between them. In the simplest binary classification scenario, if some data points are provided that are labeled into two classes representing human actions, then the goal is to find out the label or the action class of a new data point. Support vectors represent those data points that lie on the margin and closer to the hyperplane. The margin is defined to be the maximum width of the boundary from the hyperplane to the nearest data point. Given training dataset $T = \{(d_1, l1), (d_2, l2), \ldots, (d_n, l_n)\}$ with di be n-dimensional input data vector and li representing the class of input data di, where li $\in \{-1, 1\}$. The SVM classifier seeks to compute the plane $F(d)$ given by:

$$F(d) = w.d + c \tag{8.4}$$

where c is the biased coefficient and w is the weight vector indicating the hyperplane's normal. These two variables, w and c, are calculated throughout the training procedure. The $F(d)$ must give a positive value for each of the positive points of data belonging to one class and a value that is negative for the data points belonging to the other class to correctly categorize the training set T. The following may be expressed mathematically for all the data points di in T:

$$\mathbf{w.d_i} - \mathbf{c} \geq 0 \quad \text{if} \quad l_i = 1$$

$$\mathbf{w.d_i} - \mathbf{c} \leq 0 \quad \text{if} \quad l_i = -1 \tag{8.5}$$

The above conditions can be given in a single expression, for all data points $d_i \in T$, as follows:

$$l_i(\mathbf{w.d_i} - \mathbf{c}) \geq 0 \tag{8.6}$$

If the preceding linear equation is true, the data T may be separated linearly. In higher dimensions, this linear equation creates a plane that is referred to as a hyperplane. The goal is to identify the hyperplane that maximizes the margin or distance between hyperplane and the closest data points di's, because there may be several linear equations that meet Eq. (8.6). Eq. (8.6) may be expressed mathematically as follows for data points in di T:

$$l_i(\mathbf{w.d_i} - \mathbf{c}) \geq 1 \tag{8.7}$$

From any point di to the hyperplane $F(d)$, the distance is given by $\|F(d)\| / \|\mathbf{w}\|$. The distance $\|F(d)\|$ between the closest point d_i to the hyperplane $F(d)$, according to Eq. (8.6), is 1. Thus, the margin M is given by:

$$\mathbf{M} = 1/\|\mathbf{w}\| \tag{8.8}$$

Support vectors are the nearest locations di that meet Eq. (8.6). It is clear from Eq. (8.8) that the relationship between weight $\|\mathbf{w}\|$ and the margin M is inverse. Therefore, by reducing the value of w, it is possible to maximize the value of M. This finding leads to the SVM minimization issue, which is an optimization problem resolved by Karush−Kuhn−Tucker [34] utilizing Lagrange multipliers i, as follows:

$$w = \sum_{i=1}^{n} \lambda_i l_i d_i \tag{8.9}$$

When the λi's are determined along with the corresponding nearest points di representing support vectors, the biased c is calculated. Then Eq. (8.4) of classification can be expressed as:

$$\mathbf{F(d)} = \sum_{i=1}^{n} \lambda_i \, l_i \, d_i.\mathbf{d} - \mathbf{c} \tag{8.10}$$

8.3.4 Error correcting out code—based multiclass classification

DVM is used in binary classification problems to separate two classes based on the maximum separation defined by the edges in the classes. In the real world, every classification problem is basically a multivariate classification problem, not a binary classification problem. To solve multiclass information problems using SVMs, it is common practice to split the problems into binary problems and use SVM models separately. Error-correcting out code (ECOC) provides a method for using binary classifiers [35] to solve multiclass problems. The ECOC framework decomposes the problem into a binary one. Generally, ECOC is used in two stages, the encoding stage and the decoding stage. While a code word is created for each class in the coding phase, the assignment or distribution decision is made for new ideas in the decoding phase. The code word is actually a binary string of length s. The length of the code words depends on how many classifiers are trained. For each class, the classifier is trained to obtain a decision boundary that distinguishes that class from other classes. The classifier is trained on a set of functional classes (binary classes) to produce $s(s-1)/2$ decision limits. To recognize human activities in workgroups, each worker is used to predict the process of the workgroup by generating a positive response. In the proposed method, videos about human activities are passed through $s(s-1)/2$ binary classifiers, each of which is classified as positive or negative. A positive sign indicates that the action belongs to the class, and a negative sign indicates that the action does not belong to the class. Let A1, A2, A3, ... classifiers, then the total number of classifiers Ci will be $k(k-1)/2$ to classify each Ai according to the one-to-one coding scheme. For example, assuming the total number of actions is 4, the total number of objects is $4(4-1)/2 = 6$. Classifier C1 learned the data in class A1 and class A2 as positive and negative classes. Similarly, classifier C2 is trained on data with classes A1 and A3, classifier C3 is trained on data with classes A2 and A4, and so on. The whole encoding matrix is explained in Table 8.1. In Table 8.1, each column corresponds to the class and column processing class. A hexagram was obtained by examining all six workers to identify the new job during the decision-making phase. Calculate the Hamming distance and select the class that provides the smallest Hamming distance as the class action.

8.4 Existing models for securing healthcare

FC-based IoT is a popular issue right now. The data sent from healthcare IoT devices to cloud servers are often not secured and are therefore

Table 8.1 A 4-action class error-correcting out code encoding under one-versus-one scheme.

Action class	C1	C2	C3	C4	C5	C6
A1	1	1	1	0	0	0
A2	−1	0	0	1	1	0
A3	0	−1	0	−1	0	1
A4	0	0	−1	0	−1	−1

vulnerable to attack. This was not covered in the earlier articles. To the best of our knowledge, there is a risk of disclosing patient-sensitive information as a result. Additionally, it is urgent to identify healthcare IoT devices to verify and authenticate healthcare data. This can be done conveniently using blockchain in the IoT−FC system [36]. More specifically, at the edge of networks, servers should decentrally carry out some authentication and verification. In-depth discussions of current IoT, fog, and cloud technologies for the healthcare industry are provided in this section. These methods mainly concentrated on IoT device identification, IoT data authentication, and IoT cyber attacks. The following is a summary of some of the existing efforts on blockchain and healthcare IoT security:

An innovative system based on blockchain and IoT, BeeKeeper, was proposed by Zhou et al [37]. A cloud server can process the data in their suggested system by running calculations on the user data. Any node may be chosen by the current leader to serve as the server authorization leader. To implement BeeKeeper, they took advantage of the Ethereum blockchain. In order to create unique credentials, Somino et al. combined private and public attributes. The credentials can be used by users who possess these properties. The users in this case are those who use IoT devices. Nevertheless, the identification of IoT devices cannot be done using their recommended methodologies. To reduce latency and network traffic, they have combined blockchain technology with cryptographic processes in IoT. Although, IoT and blockchain scalability have not been addressed.

Similar to this, a partially centralized protocol was created by Rahulamathavan et al. The primary authority in this situation was in charge of producing the parameters for users and miners. To enable verification and decryption by some particular miners and users who hold the characteristics, the authors employed encryption-based attributes. The method helped the IoT retain secure transmission in some way. However, the issues with device identification and key authentication are not addressed in their proposed work. They mainly concentrate on centralized secure transmission. There is no distributed ecosystem.

Blockchain was employed by the creators of Ref. [38] for data integration and secure transmission. The majority of the unencrypted data are kept at the receiving site in many places using peer-to-peer file storage technology. They created protocols for cutting-edge hardware like FC, which aids end users in processing data by preserving integrity via blockchains. Applications of the consensus algorithm in proof-of-work (PoW) blockchain models are described in Ref. [39]. In which miners use a distribution of their processing capacity to solve the given task. The authors of Ref. [40] presented an FC-based architecture utilizing blockchain to increase the scalability and latency of IoT networks. The architecture is a tool for healthcare and other tele-industries to receive secure services.

The authors proposed an integrated blockchain and FC-based network architecture for the IoT in Ref. [41]. The FC nodes also offer safe and low-latency transmission to the networks for smart cities. The authors of Ref. [42] employed decentralized blockchain for Internet of Everything (IoE), enhancing the system's tamper-proof qualities. Also, they have decreased the cost of deployment, installation, and maintenance. Their suggested method protects the devices from a man-in-the-middle attack and works well for the IoE and smart cities. Moreover, blockchain stores various agreements and smart contracts for data transit and transactions. Naveed Islam et al. suggested a blockchain-based FC architecture in Ref. [43] for healthcare applications such as tracking patient activities. The SVM method is used in their proposed framework to classify and categorize the video frames based on patient activity. The approach, however, does not satisfy the need for IoT device identification in e-healthcare.

The authors of Ref. [44] suggested a blockchain-based IoE architecture for the FC environment. Blockchain fog-based architecture network is the name of the secure architecture that operates in that network (BFAN). In their suggested study, several encryption and authentication methods are used to secure data utilizing blockchain. Smart cities can use the proposed architecture. Their suggested work's major goal is to reduce latency and energy use while further enhancing secure data transmission via blockchain. Similarly, Michael Harbert et al. in Ref. [45] presented a plasma-based system to get around the problem of blockchains having a high load when used for IoT. Their suggested structure is built on the idea of fusing blockchain with FC. For the authentication and identification of IoT networks, the authors Muhammad Tahir et al. developed a novel work. The IoT network they propose is based on blockchain and employs random numbers and a probabilistic approach.

The authors presented an architecture that utilizes the PoW idea in Ref. [46]. PoW's infrastructure is compact. The main purpose of the suggested concept is to use a consensus technique to offload the computational effort to fog and cloud. They employed Stackelberg's approach, which uses computing to formulate resource management at fog nodes. The miners at the nodes are in charge of optimization. The goal of the unloading service depends on the interests of the miners. The backward induction method is used to validate and verify the model. In contrast, the authors of Ref. [47] explored several issues and research concerns related to the integration of fog and cloud for latency minimization, resource allocation, secure transmission, optimization, energy consumption, and RAM utilization. The issue of growth in heterogeneous devices and applications was raised by them. The reinforcement learning technique with computation offloading of blocks was utilized by the authors in Ref. [48]. The method operates in a distributed environment with fog nodes positioned at the edge of IoT networks. However, in their investigation, they were unable to address the problem of security and privacy for data transmission across IoT, fog, and cloud.

Similar to this, in Ref. [49], the authors presented a method for protecting cloud servers that are centrally located from outside intruders. Their suggested approach combines dynamic metadata with database schema design. In their algorithm, they have incorporated a variety of cryptographic approaches. Future projects must use a reinforcement learning algorithm to implement the suggested work. In Ref. [50], the authors used binary ATM to reduce latency, which also addressed the issue of packet error in the cloud. Their findings demonstrate that binary ATM outperforms traditional techniques. Moreover, node processing delays in binary ATMs, which rely on the number of packets, are reduced when the constant bit rate component is equal to or lower than the variable bit rate component.

The authors, however, suggested the Femto cloud approach in Ref. [51]. Their suggested approach calls for creating a mobile cloud from dynamic, self-configurable, and multidevice collection of mobile devices. With the help of this technique, numerous mobile devices can be set up in an integrated cloud computing service. The scheduler must assign tasks using tools that the scheduling algorithm may use to improve the metrics while controlling device churn. In Ref. [52], the authors presented a framework called FogBus to reduce data traffic in IoT-fog-cloud architecture by reducing network and CPU use. To reduce the heavy data traffic and protect the private, secret IoT data from outside intruders and hackers, their suggested method used a blockchain-based mechanism. To safeguard activities on sensitive IoT data, it employs a number of encryption approaches. Additionally, the suggested architecture makes it easier to integrate the IoT, fog (edge), and cloud infrastructures.

The authors also covered the use of IoT for smart healthcare applications in Ref. [53]. The IoT in healthcare, e-healthcare, and telemedicine are all topics of discussion, along with a number of related issues and potential. The authors also covered the significance of FC in reducing high latency and ensuring that time-sensitive applications in smart healthcare can achieve QoS standards. To keep track of patients' health conditions, they suggested a novel paradigm. The authors next went over several wearable technologies for monitoring and recording patient's vital signs. They built their system model using a machine-learning strategy.

In Ref. [54], the authors put out a framework for a smart healthcare system that integrates IoT and FC and uses deep learning-based methods to diagnose patient cardiac disease in real-time mode. The framework was able to satisfy the IoT for healthcare QoS criteria. It was created to function for latency-sensitive applications, including flight control and healthcare monitoring. Big data is the term for the volume of data produced by IoT in healthcare. Large computations must be performed on these data before they are transmitted to databases and then from databases to cloud data centers, which reduces system performance. Consequently, in terms of power consumption, jitter, and prediction accuracy of heart disease diagnosis, the proposed fog-enabled cloud framework satisfies the QoS for healthcare IoT.

The IoE, IoT, and medical devices that are currently in use all operate on a centralized form of communication. The cloud servers check the IoT devices, which raises the overhead associated with infrastructure and maintenance costs. IoT data can therefore be authenticated using the identity of the IoT device. Identification of healthcare IoT devices is crucial for access management and authentication. An integrated FC-based blockchain model that operates in a distributed and decentralized manner can be used to accomplish this in the absence of a single point of failure.

8.5 Discussion and analysis

It has been observed that one of the most important concerns of healthcare services is the protection and confidentiality of patient information and medical records stored by hospitals [54]. Data breaches, theft, and losses have become classics in the healthcare industry. This document is useful because it contains personal identification information [55]. In this case, it is important to ensure the confidentiality and integrity of this information. To ensure the protection and confidentiality of data, it has been recognized that it is necessary to ensure that data is not altered, lost, or stolen in such sensor problems. In this case, through the analysis of relevant documents, it has been found that the integration of blockchain technology can effectively solve this problem. Transaction blocks created in a blockchain are put together into a chain, with each block containing its own combination and the hash of the previous block. This makes it impossible to steal or modify data without getting caught. In addition, only authorized persons who have the keys can access the information. Using blockchain technology in healthcare, doctors can create a distributed yet highly secure system for their medical information. The warehouse can only be accessed by authorized personnel [56]. Thus, it solves data breach and privacy issues and also makes data easily accessible. On the other hand, FC is a distributed computing model that brings computing and data storage closer to the edge of the network, where data is generated. Fog calculates target to attack. Using fog, doctors can identify opportunities and decisions that can improve patient outcomes. Clinical trials are studies that evaluate the safety and efficacy of new treatments. Clinical trial management involves collaboration between physicians, researchers, patients, and regulators, all of whom need to share and access clinical information [57], which is very sensitive. Blockchain technology provides a secure and transparent platform for monitoring data management, while the cloud can increase the speed and efficiency of data processing and analysis [58].

8.6 Challenges in adopting blockchain technology in healthcare

- Privacy and security concerns: While blockchain and FC can offer enhanced security and privacy for healthcare data, there are still concerns around data breaches, hacking, and other security vulnerabilities.
- Integration with existing systems: Integrating blockchain and FC into existing healthcare systems can be a complex process, requiring significant changes to data management practices and IT infrastructure.
- Scalability: The scalability of blockchain and FC systems is a challenge, as these technologies require significant computing resources and can be slow to process transactions.

- Regulation and compliance: Healthcare regulations and compliance requirements can be implemented in blockchain and FC systems.
- Data standardization: Ensuring interoperability and standardization of healthcare data is critical for the success of blockchain and FC in healthcare, but achieving this can be a challenge due to the diversity of data types and formats used in healthcare.
- Cost: Implementing blockchain and FC in healthcare can be expensive, especially for small or under-resourced healthcare organizations.
- Governance and management: Blockchain and FC systems require clear governance and management frameworks to ensure their proper functioning and to prevent abuse or misuse of the technology.
- Interoperability: Interoperability between different blockchain and FC systems can be a challenge, as these systems are often proprietary and use different standards and protocols.
- Ethical and legal considerations: The use of blockchain and FC in healthcare raises ethical and legal considerations around issues such as informed consent, data ownership, and liability. It is important to ensure that these issues are properly addressed to avoid potential harm to patients and to ensure compliance with legal and regulatory frameworks.

By addressing these issues and challenges, the adoption and implementation of blockchain and FC in healthcare can be made more feasible and effective.

8.7 Future scope and issues

Blockchain technology and cloud computing have gained significant influence in the healthcare industry due to their ability to solve fundamental problems such as data privacy, security, and sharing. In the future, further efforts may be made to explore the full potential of medical technology. A potential future area of work is the development of blockchain-based electronic health records (EHR). Blockchain technology can provide a distributed, secure, and immutable platform for storing and sharing patients' medical information. This can give patients full control over their medical information, including who has access to it; moreover, the use of smart contracts on the blockchain can facilitate the exchange of health information between different healthcare providers, increase collaboration, and reduce error.

Another area of future work is the integration of blockchain technology and cloud computing to improve medical information. FC can help reduce latency and bandwidth requirements for health data by providing local processing and storage capabilities. Blockchain technology can be used to secure and verify data processing, ensuring the authenticity and integrity of the data. This can enable healthcare organizations to make more informed decisions based on real-time information, improve patient outcomes, and reduce costs. Additionally, future studies may focus on the use of blockchain technology and cloud computing in clinical trials. Clinical trials are an important part of the drug development process but are often hindered by concerns about data privacy, security, and fairness. Blockchain technology can provide a secure and immutable platform for storing and sharing medical information, while the Internet can provide local processing and storage, reducing the latency and bandwidth required for data analysis. This can increase the efficiency and effectiveness of the drug development process by allowing clinical scientists to make more informed decisions based on real-time data. Finally, future studies may explore blockchain technology and cloud computing applications in telemedicine. Telemedicine has grown in popularity in recent years, but it is often thwarted due to concerns about data privacy, security, and collaboration. Blockchain technology can provide a secure and transparent platform for storing and sharing information, while the Internet can provide local processing and storage, reducing the low latency and bandwidth required for data analysis. This can improve access to healthcare and reduce costs by enabling doctors to provide quality care to patients wherever they are. In conclusion, blockchain technology and cloud computing have the potential to revolutionize the healthcare industry by solving fundamental problems such as data privacy, security, and communication. Future work may focus on exploring the full potential of this technology in areas such as EHR, clinical data analysis, clinical trials, and telemedicine. These efforts help improve patient outcomes, reduce costs, and streamline drug development, resulting in better health and outcomes.

8.8 Conclusion

The current trend in healthcare is mostly reflected in the digitization of data. Where smart contracts will play an important role. IoT data require stable operation and maintenance. Medical IoT creates diversity and true Patient Health Data (PHD). Processing large amounts of data can lead to seamless transfer between IoT devices and users. Therefore, in this article, we cover the issue of medical IoT authentication and identification of

IoT devices. Secondly, we provide a fresh approach to solve the PHD by regular instance central task filter PHD in a more reliable manner. These central servers are prone to a single failure. Additionally, healthcare IoT devices can be vulnerable to outside intruders, hackers, and malicious actors. This can cause data corruption. A large number of heterogeneous IoT devices make PHD unreliable and unproven. We, therefore, proposed a three-tier FC-based blockchain architecture, a mathematical framework, an advanced signature-based encryption algorithm (ASE), and an analytical model for healthcare IoT device identification and PHD authentication for secure healthcare IoT data transmission to address the aforementioned issue. The solution and implementation of the proposed work are conducted in iFogSim, SimBlock, and Python Editor Tool. A proposed algorithm improves the detection accuracy and reliability of malware; also, the ASE algorithm reduces packet errors for PHD transmission between medical IoT and end users. When the performance evaluation and evaluation results are compared, the proposed system easily outperforms existing systems and systems such as Sis Bus, Beekeeper, Femto cloud, and FBAN.

Implementation of the proposed model and PHD authentication and medical IoT device identification are considered as a solution process, reducing the percentage of packet failure and bad disease. The plan has a greater impact on the success and duration of work on healthcare IoT and FC because doctors' time can be spread over minimally serviced patients and physicians in a single calculation. Secure medical information is generated at the edge of IoT network in a decentralized way by using blockchain technology. Future work includes testing algorithms to reduce the complexity of medical IoT−FC systems as the number of IoT and cloud systems increases. This plan can also be used for telesurgery and augmented reality. In addition, the proposed method can be used in other IoT applications such as offshore oil and gas monitoring in the future. In future research, we will test the ASE algorithm to overcome the scalability limitations of blockchain when used in conjunction with medical IoT and FC.

References

[1] K. Biswas, V. Muthukkumarasamy, "Securing smart cities utilising blockchain technology," in: Proceedings of the 18th IEEE International Conference on High Performance Computing and Communications,14th IEEE International Conference

of Smart City, 2nd IEEE International Conference on Data Science and Systems HPCC/SmartCity/DSS, December 2016, pp. 1392−1393.

[2] S. Karale, V. Ranaware, Blockchain technology applications in smart city development: a research, International Journal of Innovative Technology and Exploratory Engineering 8 (11) (2019) 556−559.

[3] Y. Zhang, M.M. Kamruzzaman, and L. Feng, Complex system of vertical baduanjin lifting motion sensing recognition against the backdrop of big data, Complexity, 2021, article ID 6690606 (1−10), 2021.

[4] S. Qian, T. Zhang, C. Xu, M.S. Hossain, Social event classification using boosted multimodal supervised latent dirichlet allocation, ACM Transactions on Multimedia Computing, Communications, and Applications 11 (2) (2014) 1−27.

[5] A.K. Sangaiah, D.V. Medhane, G.-B. Bian, A. Ghoneim, M. Alrashoud, M.S. Hossain, Energy-Aware green adversary model for cyberphysical security in industrial systems, IEEE Transactions on Industrial Informatics 16 (5) (2020) 3322−3329.

[6] S.A. Alanazi, M.M. Kamruzzaman, M.N. Islam Sarker, et al., Boosting breast cancer diagnosis using convolutional neural network, Journal of Healthcare Engineering 2021 (2021) 1−11.

[7] B. Yuan, M.M. Kamruzzaman, S. Shan, Application of motion sensor based on neural network in basketball technology and physical fitness evaluation system, Wireless Communications and Mobile Computing 2021 (2021) 5562954.

[8] Y. Xu, M. Wei, M.M. Kamruzzaman, Inter/intra-category discriminative features for aerial image classification: a quality-aware selection approach, Next Generation Computer Systems 119 (2021) 77−83. vol.

[9] C. Esposito, M. Ficco, B.B. Gupta, Blockchain-based authentication and authorization for smart city applications, Information Processing & Management 58 (2) (2021) 102468.

[10] Y. Chen, S. Das, P. Dhar, et al., Detecting and mitigating IP − faked widespread DoS attacks, International Journal on Network Security 7 (1) (2008) 70−81.

[11] Y. Shi, S. Wang, S. Zhou, M.M. Kamruzzaman, Research on modelling approach of forest tree image recognition based on CCD and theodolite, IEEE Access 8 (2020) 159067−159076.

[12] M.S. Hossain, G. Muhammad, N. Guizani, et al., AI and mass surveillance system-based healthcare framework to battle COVID-I9 type pandemics, IEEE network 34 (4) (2020) 126−132.

[13] X. Li, J. Zhong, M.M. Kamruzzaman, Complex robot activity recognition by quality-aware deep reinforcement learning, Future Generation Computer Systems 117 (2021) 480−485.

[14] M.M. Kamruzzaman, Arabic sign language detection and Arabic voice generation with convolutional neural networks, Wireless Communications and Mobile Computing 2020 (2020).

[15] H. Treiblmaier, A. Rejeb, A. Strebinger, Blockchain as a driver for smart city development: application sectors and a comprehensive research programme.", Smart Cities 3 (3) (2020) 853−872.

[16] B. Burnes, N. Towers, Consumers, clothes shops, and production planning and control in the smart city, Production Planning & Control 27 (6) (2016) 490−499.

[17] M.K. Al-Azzam, M.B. Al-Azzam, Smart city and smart-health framework, problems and opportunities, International Journal of Advanced Computer Science and Applications 10 (2) (2019) 171−176.

[18] M.M. Kamruzzaman, Architecture of smart healthcare system utilising AI, in: Proceedings of the 2020 IEEE International Conference on Multimedia & Expo Workshops (ICMEW), July 2020, pp. 1−6, IEEE, London, UK.

[19] M.N. Islam Sarker, M.M. Kamruzzaman, Md Enamul Huq, R. Zaman, et al., Smart city governance through big data: transformation towards sustainability, in: Proceedings of the 2021 International Conference of Women in Data Science at Taif University, WiDSTaif, March 2021, pp. 1–6, IEEE, Taif, Saudi Arabia.

[20] S.A. Alanazi, M.M. Kamruzzaman, M. Alruwaili, N. Alshammari, S.A. Alqahtani, A. Karime, Monitoring and avoiding COVID-19 using the SIR model and machine learning in smart health care, Journal of Healthcare Engineering 2020 (2020) 8857346.

[21] A.D. Dwivedi, G. Srivastava, S. Dhar, R. Singh, A decentralised privacy-preserving healthcare blockchain for IoT, Sensors (Basel, Switzerland) 19 (2) (2019) 1–17.

[22] M.A. Rahman, M.S. Hossain, G. Loukas, et al., Blockchain-based mobile edge computing platform for secure treatment applications, IEEE Access 6 (2018) 72469–72478. vol.

[23] S.K. Lo, X. Xu, Y.K. Chiam, Q. Lu, "Evaluating applicability of applying blockchain," in: Proceedings of IEEE International Conference on Engineering Complex Computer Systems, vol. 2017, pp. 158–161, 2018.

[24] M.A. Rahman, E. Hassanain, M.M. Rashid, S.J. Barnes, M.S. Hossain, Spatial blockchain-based secure mass screening framework for children with dyslexia, IEEE Access 6 (2018) 61876–61885. vol.

[25] Z. Zheng, S. Xie, H. Dai, X. Chen, H. Wang, An overview of blockchain technology: architecture, consensus, and future trends, in: Proceedings of the 2017 IEEE 6th International Conference on Big Data, BigData Congr., June 2017, pp. 557–564, IEEE, Honolulu, HI, USA.

[26] D. Yaga, P. Mell, N. Roby, K. Scarfone, Blockchain Technology Overview, NIST, Gaithersburg, MD, 2018.

[27] L. Atzori, A. Iera, G. Morabito, Understanding the Internet of Things: definition, potentials, and societal role of a fast evolving paradigm, Ad Hoc Networks 56 (2017) 122–140. vol.

[28] M.S. Hossain, S.U. Amin, G. Muhammad, M. Alsulaiman, . Using deep learning for epileptic seizure detection and brain mapping visualisation, ACM Transactions on Multimodal Computing, Communications, and Applications 15 (1s) (2019) 1–17.

[29] A. Tewari, B.B. Gupta, Security, privacy, and trust of distinct layers in Internet-of-Things (IoTs) architecture, Next Generation Computer Systems 108 (2020) 909–920. vol.

[30] S.B. Baker, W. Xiang, I. Atkinson, Internet of Things for smart healthcare: technology, problems, and possibilities, IEEE Access 5 (c) (2017) 26521–26544.

[31] J. MacQueen, Some methods for classification and analysis of multi- variate observations, in: Proceedings of the Fifth Berkeley Symposium on Mathematical Statistics and Probability, in: Statistics, vol. 1, University of California Press, Berkeley, Calif., 1967, pp. 281–297. Available from: <https://projecteuclid.org/euclid.bsmsp/1200512992>.

[32] C. Cortes, V. Vapnik, Support-vector networks, Machine Learning 20 (3) (1995) 273–297.

[33] M.C. Ferris, T.S. Munson, Interior-point methods for massive support vector machines, SIAM Journal on Optimization 13 (3) (2002) 783–804.

[34] T.G. Dietterich, G. Bakiri, Solving multiclass learning problems via error-correcting output codes, Journal of Artificial Intelligence Research 2 (1) (1995) 263–286.

[35] K. Abouelmehdi, A. Beni-Hessane, H. Khaloufi, Big healthcare data: preserving security and privacy, Journal of Big Data 5 (1) (2018) 1–18.

[36] W. She, Q. Liu, Z. Tian, J.-S. Chen, B. Wang, W. Liu, Blockchain trust model for malicious node detection in wireless sensor networks, IEEE Access 7 (2019) 38947–38956.

[37] L. Zhou, L. Wang, Y. Sun, P. Lv, Beekeeper: a blockchain-based iot system with secure storage and homomorphic computation, IEEE Access 6 (2018) 43472−43488.

[38] L. Wu, X. Du, W. Wang, B. Lin, An out-of-band authentication scheme for Internet of Things using blockchain technology 2018 International Conference on Computing, Networking and Communications (ICNC), IEEE (2018) 769−773.

[39] A. Ouaddah, A. AbouElkalam, A.A. Ouahman, Towards a Novel Privacy-Preserving Access Control Model Based on Blockchain Technology in IoT Europe and MENA Cooperation Advances in Information and Communication Technologies, Springer, 2017, pp. 523−533.

[40] T. Alam, IoT-Fog: a communication framework using blockchain in the Internet of Things, arXiv preprint arXiv:1904.00226, 2019.

[41] K. Singh, O. Dib, C. Huyart, K. Toumi, A novel credential protocol for protecting personal attributes in blockchain, Computers & Electrical Engineering 83 (2020). Article 106586.

[42] B. Bhushan, A. Khamparia, K.M. Sagayam, S.K. Sharma, M.A. Ahad, N.C. Debnath, Blockchain for smart cities: a review of architectures, integration trends and future research directions, Sustainable Cities and Society 61 (2020). Article 102360.

[43] N. Islam, Y. Faheem, I.U. Din, M. Talha, M. Guizani, M. Khalil, A blockchain-based fog computing framework for activity recognition as an application to e-Healthcare services, Future Generation Computer Systems 100 (2019) 569−578.

[44] P. Singh, A. Nayyar, A. Kaur, U. Ghosh, Blockchain and fog based architecture for internet of everything in smart cities, Future Internet 12 (4) (2020) 61.

[45] M.H. Ziegler, M. Großmann, U.R. Krieger, Integration of fog computing and blockchain technology using the plasma framework 2019 IEEE International Conference on Blockchain and Cryptocurrency (ICBC), IEEE (2019) 120−123.

[46] L. Chen, Z. Wang, F. Li, Y. Guo, K. Geng, A. Stackelberg, Security game for adversarial outbreak detection in the Internet of Things, Sensors (Basel, Switzerland) 20 (3) (2020) 804.

[47] L. Bittencourt, et al., The Internet of Things, fog and cloud continuum: integration and challenges, Internet of Things 3−4 (2018) 134−155.

[48] M.G.R. Alam, Y.K. Tun, C.S. Hong, Offloading code using multi-agent and rein-forcement learning in mobile fog 2016 International Conference on Information Networking (ICOIN), IEEE (2016) 285−290.

[49] K. Sambyo, C.T. Bhunia, Improved performance of integrated phone, video, and data services with the usage of multi level ATM to decrease latency in clouds, in: 2014 11th International Conference on Information Technology: New Generations, IEEE, 2014, pp. 607−607.

[50] K. Habak, M. Ammar, K.A. Harras, E. Zegura, Femto clouds: leveraging mobile devices to provide cloud service at the edge 2015 IEEE 8th international conference on cloud computing, IEEE (2015) 9−16.

[51] S. Tuli, R. Mahmud, S. Tuli, R. Buyya, FogBus: a blockchain-based lightweight framework for edge and fog computing, Journal of Systems and Software 154 (2019) 22−36.

[52] S.B. Baker, W. Xiang, Internet of things for smart healthcare: technologies, chal-lenges, and opportunities, IEEE Access 5 (2017) 26521−26544.

[53] S. Tuli, et al., An ensemble deep learning-based smart healthcare system called Healthfog uses integrated IoT and fog computing environments to automatically diagnose cardiac problems, Future Generation Computer Systems 104 (2020).

[54] Y. Wang, A. Kogan, Developing privacy-preserving Blockchain-based transaction processing systems, International Journal of Accounting Information Systems 30 (2018) 1−18. vol.

[55] R. El-Gazzar, K. Stendal, Blockchain in healthcare: hope or hype? Journal of Medical Internet Research 22 (7) (2020) e17199.

[56] F. Bonomi, R. Milito, J. Zhu, S. Addepalli, Fog computing and its function in the Internet of Things, in: Proceedings of the first MCC Workshop on Mobile Cloud Computing — MCC '12, p. 13, ACM, New York, NY, USA, August 2012.

[57] D. Wu, S. Liu, L. Zhang, et al., A fog computing-based framework for process monitoring and prognosis in cyber-manufacturing, Journal of Manufacturing Systems 43 (2017) 25—34. vol.

[58] H. Bay, A. Ess, T. Tuytelaars, L.V. Gool, Speeded-up robust features (SURF), Computer Vision and Image Understanding 110 (3) (2008) 346—359. Similarity Matching in Computer Vision and Multimedia.

CHAPTER 9

Forensics in the Internet of Medical Things

Ankit Garg[1,2], Anuj Kumar Singh[3], A. Mohit[1,2] and A. Aleem[4]
[1]AIT-CSE, Chandigarh University, Mohali, Punjab, India
[2]University Centre for Research & Development (UCRD), Chandigarh University, Mohali, Punjab, India
[3]Amity University, Gwalior, Madhya Pradesh, India
[4]Department of CSE, UIE, Chandigarh University, Mohali, Punjab, India

9.1 Introduction

The Internet of Medical Things (IOMT) has transformed healthcare by integrating medical equipment, wearables, and health-care systems, enabling remote patient monitoring, real-time data analysis, and personalized treatment. However, as IOMT systems become more complicated and linked, additional issues emerge, notably in protecting the integrity, security, and privacy of sensitive medical data. In tackling these problems, forensic investigation plays a critical role by offering a methodical and scientific methodology to analyzing digital data, identifying potential breaches, and recreating occurrences in IOMT settings. This chapter covers the importance of forensic investigation in the context of IOMT, as well as essential ideas, problems, and approaches, and emphasizes the need for more study and cooperation to develop the area of IOMT as shown in Fig. 9.1.

9.1.1 Background and significance of Internet of Medical Things

The interconnected network of medical devices, sensors, wearables, and health-care systems that gather, monitor, and send health-related data is referred to as the IOMT [1]. The widespread availability of IOMT devices has transformed the health-care business, allowing for remote patient monitoring, real-time data analysis, and personalized health-care delivery [2]. This networked ecosystem has enormous promise for improving patient outcomes, increasing health-care efficiency, and lowering costs [3]. IOMT devices enable health-care practitioners to provide proactive and personalized treatment by continuously monitoring vital signs, tracking medication adherence, and aiding early diagnosis of health concerns [4]. The integration of IOMT with electronic health records

Securing Next-Generation Connected Healthcare Systems
DOI: https://doi.org/10.1016/B978-0-443-13951-2.00007-6

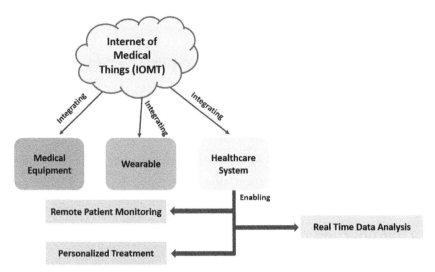

Figure 9.1 Internet of Medical Things and its need.

(EHRs) and telemedicine systems allows for even more seamless data exchange, remote health-care services, and informed decision-making [5]. However, the fast deployment of IOMT devices raises security, privacy, and data integrity problems [6]. Cyberattacks, data breaches, and unauthorized access are all increased when sensitive patient data are sent and stored inside networked systems [7]. To retain confidence in these technologies and preserve sensitive health information, it is critical to secure patient privacy and ensure the security of IOMT systems [8]. In the realm of IOMT, forensic investigation plays an important role in tackling these difficulties by offering a methodical and scientific methodology to analyzing digital data, identifying probable breaches, and recreating occurrences [9]. Health-care organizations and law enforcement agencies may limit threats, protect patient data, and maintain the integrity of IOMT systems by performing extensive forensic investigations [10].

This chapter digs into the subject of forensics in IOMT, offering insights into the particular problems and concerns involved in conducting forensic analysis of linked medical devices. Further, the chapter explores many forms of digital evidence discovered in IOMT, the forensic investigation process in IOMT environments, specialized forensic techniques and tools, and the legal and ethical ramifications. This chapter discusses the improvements of forensics in IOMT and also provides different future research directions in healthcare.

9.1.2 Overview of the increasing use of connected medical devices in healthcare

The health-care sector is seeing a fast growth of connected medical devices, which is being driven by technological improvements and the demand for more efficient and patient-centric health-care delivery. Wearable gadgets, implanted sensors, remote monitoring systems, and smart health-care infrastructure are examples of these devices [11]. In health-care settings, connected medical devices have several advantages. They allow for real-time monitoring of patients' vital signs, allowing health-care practitioners to continually track and analyze health data [12]. This continuous monitoring allows for the early diagnosis of medical issues, prompt intervention in crises, and personalized treatment programs tailored to specific patient needs [13]. Connected device integration with health-care systems and EHRs improves care coordination and streamlines data management [14]. The seamless exchange of patient data between devices and health-care providers allows for full patient profiles, avoids test duplication, and promotes evidence-based decision-making [15]. Furthermore, networked medical devices enable people to take an active role in their own treatment. Individuals may monitor their health metrics, track their progress, and obtain personalized health suggestions using wearable devices and smartphone apps. This generates a sense of empowerment and participation, which leads to improved health outcomes and patient satisfaction. However, the rising proliferation of linked medical equipment raises cybersecurity concerns. Because these devices are linked and send sensitive health data, they are vulnerable to cyberattacks, data breaches, and unauthorized access [16]. Malicious actors may exploit device vulnerabilities or intercept transmitted data, jeopardizing patient privacy, and the integrity of health-care systems. In this setting the forensics sector in IOMT becomes critical (see Fig. 9.2). Forensic investigation is critical to protecting the security and integrity of connected medical equipment, as well as limiting the risks associated with cybersecurity attacks.

Health-care organizations can detect and respond to security issues, identify the source of breaches, and adopt preventative steps to secure patient data by undertaking rigorous forensic investigation [17]. As a result, knowing the growing usage of linked medical devices in healthcare is critical to grasping the urgency and significance of forensic investigation in IOMT. This chapter seeks to offer an overview of the connected medical device ecosystem, emphasizing the importance of strong forensic practices to secure these devices and preserve patient privacy in the face of emerging cybersecurity threats.

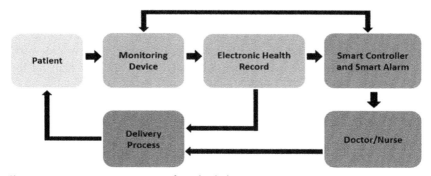

Figure 9.2 Forensics in Internet of Medical Things.

9.1.3 Importance of forensic investigation in ensuring the integrity and security of Internet of Medical Things

The function of forensic investigation in assuring the integrity and security of IOMT systems is critical. As the health-care sector increasingly depends on linked medical devices and the transfer of sensitive patient data, identifying and addressing possible security breaches, protecting patient privacy, and maintaining trust in these technologies become critical. The expanding threat landscape is one of the primary reasons for the necessity of forensic investigation in IOMT. Cyberattacks on health-care organizations and related medical equipment have grown in sophistication and frequency. Malicious actors may attempt to exploit IOMT device vulnerabilities, get unauthorized access to patient data, or disrupt vital health-care infrastructure. Forensic investigation assists health-care organizations in detecting and mitigating risks, allowing them to respond efficiently and secure their systems and patients. Furthermore, forensic investigation is critical in proving culpability and attribution. In the case of a security incident or data breach, forensic analysis can assist in determining the source of the breach, the degree of the harm, and the responsible party. This information is critical for legal procedures, internal investigations, and instituting preventative measures.

Another critical part of forensic inquiry in IOMT is preserving the integrity of digital data. Sensor data, device logs, and network traffic can all give useful insights into the sequence of events that led to a security incident or data breach. Forensic investigators can recreate the chain of events, identify possible vulnerabilities, and apply measures to improve the security and integrity of IOMT systems by storing and analyzing these data. Furthermore, forensic investigation in IOMT settings aids in regulatory compliance and industry standard adherence. Various data protection and privacy legislation apply to health-care organizations, including the Health Insurance Portability and Accountability Act (HIPAA) and the General Data Protection Regulation (GDPR) [18]. Organizations can demonstrate their commitment to complying with these rules and preserving patient information by undertaking forensic investigations. Overall, forensic investigation is critical to the integrity and security of IOMT systems. Forensic investigation contributes to the trustworthiness and reliability of connected medical devices, protects patient privacy, and improves overall health-care cybersecurity by detecting and mitigating security threats, establishing accountability, preserving digital evidence, and ensuring regulatory compliance.

9.2 Overview of forensics in Internet of Medical Things

Forensics in the IOMT is the analysis of digital evidence in linked medical devices, wearables, and health-care systems using investigative techniques. It is critical in maintaining the integrity, security, and privacy of IOMT systems, as well as identifying and investigating possible events or breaches. This chapter gives an introduction of the discipline, emphasizing major ideas such as forensic investigative techniques, digital evidence types encountered, obstacles, and the use of artificial intelligence (AI) and machine learning (ML). It emphasizes the need of evidence preservation, regulatory compliance, and ethical issues while also investigating emerging trends and research prospects for developments in IOMT forensics.

9.2.1 Definition and scope of forensic investigation in the context of Internet of Medical Things

In the context of the IOMT, forensic investigation refers to the systematic and rigorous process of collecting, analyzing, and interpreting digital evidence to uncover potential security breaches, identify incident root causes, and establish accountability in the realm of connected medical devices. In IOMT the scope of forensic inquiry includes numerous components of the interrelated ecosystem. It entails investigating the security and integrity of connected medical devices' hardware, software, and communication protocols. This involves examining the design and implementation of device firmware, operating systems, authentication methods, and encryption algorithms for potential flaws that attackers may exploit. In addition, forensic investigation in IOMT includes the examination of communication channels and protocols used to transfer data between devices, wearables, and health-care systems. The forensic investigation process in IOMT can be improved by examining network traffic, data records, and metadata for unauthorized access, data tampering, or malicious activity. It also includes evaluating the performance of the IOMT infrastructure's data encryption, integrity checks, and access control measures.

Fig. 9.3 showcases the key steps that are involved in the forensic investigation process in IOMT. The detail of each step is given next.

- Identifying and preserving digital evidence: This involves identifying prospective digital evidence sources such as device logs, sensor data, network traffic, and patient health information. To maintain its integrity and acceptability in judicial processes, evidence must be collected and stored using suitable methodologies [19].
- Forensic analysis: The digital evidence is rigorously analyzed utilizing specialized tools and methodologies. The process of forensic analysis can be imporved by examining the evidence for signs of compromise, abnormal behavior, or unauthorized access. Reconstructing the

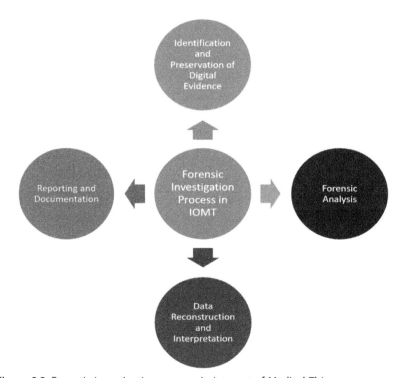

Figure 9.3 Forensic investigation process in Internet of Medical Things.

chronology of events leading up to the incident and identifying possible vulnerabilities or security holes are also part of the process [20].

- Data reconstruction and interpretation: The forensic investigator reconstructs the sequence of events by analyzing evidence and connecting various data sources. This aids in comprehending the nature of the occurrence, establishing the impact on patient safety and privacy, and identifying the people accountable [21].

- Reporting and documentation: The forensic investigation's results are documented in a complete report that contains a synopsis of the occurrence, the methodology used, the evidence gathered, and the analysis performed. The report may also contain remedial recommendations, security control changes, and strategies to prevent repeat events [22].

Fig. 9.3 depicts the forensic investigation procedure in IOMT. In IOMT, forensic inquiry expands beyond the assessment of individual devices to include the entire ecosystem, which includes wearables, medical implants, health-care systems, and data storage infrastructure. It necessitates interdisciplinary knowledge, including cybersecurity, digital forensics, health-care legislation, and legal concerns. Health-care organizations and forensic experts can effectively identify and address security breaches, mitigate risks, protect patient privacy, and improve the overall security and trustworthiness of connected medical devices and the health-care ecosystem as a whole by conducting forensic investigations in the context of IOMT.

9.2.2 Challenges and unique considerations in conducting forensic analysis of connected medical devices

Due to the complexity of the devices and the criticality of the health-care setting, forensic investigation of linked medical devices poses various obstacles and specific concerns. The following are the major issues and concerns while doing forensic investigation on linked medical devices:

- Heterogeneity of devices: Connected medical devices exist in a variety of forms, from implanted devices to wearable sensors and medical equipment. Hardware designs, operating systems, and data storage techniques may differ among devices. To successfully analyze a wide range of devices and their specific forensic problems, forensic investigators must have a thorough grasp of them.

- Analysis of real-time data: Connected medical devices create real-time data that are crucial for patient monitoring and health-care decision-making. It is difficult to analyze these data in a timely manner without interfering with the device's performance [23]. Forensic investigators must create procedures that enable the quick processing of real-time data while assuring the correctness and dependability of the forensic results.

- Regulatory compliance: Connected medical devices must comply with severe regulatory standards such as data protection, privacy, and medical device laws. To protect the admissibility and integrity of the evidence, forensic investigators must follow certain standards. During the forensic analysis process, compliance with standards such as ISO 27001, IEC 62304, and FDA recommendations is critical.
- Time-sensitive investigations: Due to the crucial nature of health-care services, forensic investigations involving linked medical equipment frequently necessitate a quick response. Rapid detection and control of security issues is critical to protect patients and the health-care system. Forensic investigators must be able to perform time-sensitive investigations without sacrificing analytical quality.
- Collaboration and expertise: Forensic examination of linked medical devices frequently necessitates collaboration among multiple stakeholders, such as forensic specialists, health-care practitioners, device manufacturers, and legal authorities. These investigations need skills in cybersecurity, digital forensics, health-care laws, and medical device technology due to their interdisciplinary nature.

To address these obstacles and special concerns, ongoing research and development in the field of forensic analysis of linked medical devices is required. By addressing these challenges, forensic investigators will be able to successfully identify security breaches, establish accountability, and contribute to the overall security and trustworthiness of connected medical devices in the health-care domain.

9.2.3 Importance of preserving digital evidence in Internet of Medical Things environments

In the context of the IOMT, preserving digital evidence is critical for conducting successful forensic investigations and protecting the integrity and security of linked medical equipment. The following are some of the most important reasons for protecting digital evidence in IOMT environments:

- Establishing credibility and validity: In forensic investigations, digital evidence acts as a trustworthy and dependable source of information. Preserving digital evidence from linked medical equipment contributes to the results' reliability and validity. It keeps a reliable record of events, activities, and interactions, which can be used in legal procedures, internal investigations, and incident response operations [24].

- Identifying the root cause of incidents: By preserving digital evidence, forensic investigators may recreate the chain of events that led to security incidents or data breaches. Investigators can discover the fundamental cause of the incident, such as unauthorized access, malware infection, or system misconfiguration, by analyzing the saved evidence. Understanding the fundamental cause is critical for adopting suitable corrective actions and avoiding future accidents [25].
- Detecting insider threats and malicious actions: Connected medical devices are susceptible to insider threats or malicious actions carried out by workers, contractors, or authorized persons. Investigators can uncover suspicious behaviors by analyzing user actions, system logs, and network traffic if digital proof is preserved. This aids in the detection and mitigation of insider threats, the integrity of IOMT systems, and the protection of patient data.
- Ensure regulation and standard compliance: Preserving digital evidence is critical for establishing compliance with health-care legislation and industry standards. Regulations such as HIPAA, GDPR, and medical device—specific requirements must be followed by health-care organizations. Evidence that has been preserved can be used to validate compliance efforts, exhibit due diligence, and establish that proper security controls and privacy protections were in place [26].
- Increasing trust and confidence: Preserving digital evidence in IOMT contexts increases trust and confidence in linked medical devices and the health-care system as a whole. It displays a dedication to security, privacy, and transparency. Health-care organizations may retain patient confidence, secure sensitive information, and encourage the uptake and utilization of connected medical devices for enhanced patient care by successfully preserving digital evidence [27,28].

9.3 Types of digital evidence in Internet of Medical Things

During forensic investigations in the sphere of the IOMT, numerous sorts of digital evidence may be uncovered.

Sensor data created by medical devices and wearables, health-related information kept in EHRs, metadata linked with IOMT devices and communications, and log files documenting system operations are all examples of this as shown in Fig. 9.4. Forensic analysts can use these many types of digital evidence to recreate events, discover abnormalities,

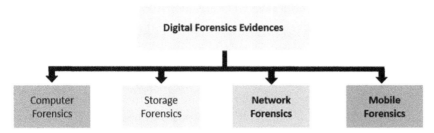

Figure 9.4 Digital forensics evidences.

and detect possible security breaches. Understanding and successfully using digital evidence is critical for guaranteeing the integrity and dependability of forensic investigations in IOMT systems.

9.3.1 Examination of different types of digital evidence encountered in Internet of Medical Things systems

In the context of the IOMT, forensic investigations meet many forms of digital evidence that are critical for comprehending security incidents, detecting vulnerabilities, and establishing the sequence of events. The assessment of various forms of digital evidence in IOMT systems necessitates the use of specialized procedures and instruments. The following are some of the most common forms of digital evidence discovered in IOMT systems:

- Device logs: Logs generated by connected medical devices capture different system operations, events, and faults. These logs include useful information regarding device operations, user interactions, and any abnormalities. Examining device logs can aid in the detection of unauthorized access attempts, system faults, or strange behavior that could signal a security compromise.

- Network traffic: For data transfer between devices, sensors, and healthcare systems, IOMT systems rely on network connectivity. Analyzing network traffic reveals information about data flow, communication patterns, and potential security flaws. Forensic investigators can uncover suspicious connections, unauthorized access attempts, or malicious actions within the IOMT infrastructure by analyzing network traffic [29–31].

- Sensor data: Sensor data, including vital signs, patient biometrics, and ambient characteristics, is frequently collected and sent by connected medical equipment. Sensor data may be used to determine the status of the device, patient circumstances, and probable data manipulations in forensic investigations. Investigators can confirm the correctness and integrity of acquired data by examining sensor data [32,33].

- Device firmware: The firmware of linked medical devices comprises important instructions and software that manage device functioning. Examining device firmware is critical for identifying any vulnerabilities, backdoors, or malicious programs that might jeopardize the device's security and integrity. It entails analyzing the firmware image, reverse engineering, and code inspection to identify potential security vulnerabilities.
- EHR: In IOMT settings, EHRs contain important information about patients, their medical histories, and treatment plans. Examining EHRs can reveal possible privacy violations, unauthorized access to patient records, or data manipulation. Investigators examine EHRs to see whether patient data have been hacked and to identify the scope and effect of any security problems.
- Communication protocols: For data sharing, IOMT systems rely on several communication protocols such as Bluetooth, Wi-Fi, or cellular networks. Analyzing protocol-specific data packets, encryption algorithms, and possible vulnerabilities are all part of the protocol examination process. It aids in identifying communication channel flaws and potential points of unauthorized access or data interception.

Forensic investigators use specialized forensic tools, data analysis methodologies, and experience in cybersecurity and medical device technology to examine various sorts of digital evidence discovered in IOMT systems. To improve the security and integrity of IOMT systems, the objective is to extract important information, establish the chain of events, identify potential security breaches, and assist evidence-based decision-making.

9.3.2 Sensor data and health-related information as potential evidence sources

Within the IOMT ecosystem, sensor data and health-related information play an important role as prospective evidence sources in forensic investigations. These sites offer useful information on patient health, device functionality, and potential security problems. As evidence, analyzing sensor data and health-related information can help to understand the context, effect, and root causes of occurrences in IOMT systems. The following expands on their significance:

- Vital signs and biometric data: Connected medical equipment frequently records vital signs and biometric data, such as heart rate, blood pressure, temperature, and oxygen levels. These data points are critical evidence sources in forensic investigations, allowing investigators to determine the patient's physiological status at a given moment. Vital sign changes or abnormalities can suggest tampering, system faults, or unauthorized access to the device [34].

- Event-triggered sensor data: Many IOMT devices have event-triggered sensors that record certain events or occurrences, such as drug delivery, device activation, or device alerts. This sort of sensor data may be used to determine the sequence of events that led to an incident or the activities performed by users. Analyzing event-triggered sensor data aids in the reconstruction of the chronology and the detection of any deviations or unauthorized activity.
- Device configuration and settings: Sensor data and health-related information might reveal device configuration and settings during an occurrence. Analyzing these data enables investigators to identify whether the gadget was operating within its boundaries or whether unauthorized adjustments were made. Device configuration or settings changes can aid in the detection of potential security breaches, criminal actions, or user mistakes.
- Patient-device interaction: Sensor data and health-related information can give insight on how patients engage with connected medical devices. Examining these data can indicate how patients engaged with the devices, such as device usage patterns, input instructions, or user modifications. It assists investigators in comprehending user behavior, probable user mistakes, or purposeful activities that may have resulted in security issues [35].
- Correlation with other digital evidence: Sensor data and health-related data can be linked to other forms of digital evidence found in IOMT systems, such as device logs, network traffic, or authentication records. This correlation helps investigators to build a complete picture of the occurrence, assess the integrity of the evidence, and find possible linkages between various pieces of evidence.

Forensic investigators address considerations like data accuracy, data integrity, data gathering procedures, and the possible influence on patient privacy when analyzing sensor data and health-related information as evidence in IOMT systems. Investigators can acquire insights into the occurrence, causes, and repercussions of security events by leveraging various evidence sources, resulting to improved security practices, improved patient safety, and stronger IOMT system resilience.

9.3.3 Metadata and log files as valuable sources for forensic analysis

Within the framework of the IOMT, metadata and log files are significant sources for forensic research. These sources include critical information regarding the creation, modification, and the use of digital assets, revealing system activities, user interactions, and possible security events. Forensic investigators can recreate events, establish timelines, and uncover significant trends by analyzing metadata and log files. The following goes into further detail on the significance of metadata and log files in forensic analysis:

- Metadata: Metadata is descriptive data about digital assets such as files, documents, or data records. It includes information such as the creation and modification times, file size, authorship, and access rights. Metadata analysis aids in establishing the authenticity, integrity, and ownership of digital assets inside IOMT systems. Investigators can use it to detect when files were generated or updated, track user actions, and establish linkages between various entities.

- File access logs: Log files keep track of file access actions such as read, write, delete, and modification. These logs record information about each access event, such as the user ID, date, time, and IP address. Investigators can trace user behaviors, detect unauthorized access attempts, and create a chronology of file-related events by reviewing file access logs. It aids in assessing whether sensitive data were read or updated without permission [36].

- System logs: System logs keep a detailed record of all system-level actions, events, faults, and warnings. These logs record data on device activities, network connections, software installations, and system changes. Analyzing system logs enables investigators to recreate the chain of events that led to an incident, uncover system vulnerabilities or misconfigurations, and monitor possible attackers' activity. System logs are essential for identifying unauthorized access, system breaches, and aberrant activity in IOMT systems.

- Network traffic logs: Data transmissions, network connections, and communication patterns between devices and systems are all recorded in network traffic logs. These logs provide information such as source and destination IP addresses, protocols, packet sizes, and timestamps. Analyzing network traffic records assists investigators in detecting unusual network activity, unauthorized connections, and probable data exfiltration efforts. It provides network-based event reconstruction and aids in determining the root cause of security breaches.

- Authentication and user activity logs: Authentication and user activity logs in IOMT systems record information on user login events, session lengths, and user-initiated actions. These logs give a trace of user activity such as user access, privilege escalation, and permission modifications. Investigators can trace user behaviors, identify unauthorized account usage, and create a chronology of user-related events by analyzing authentication and user activity logs. It aids in establishing particular user involvement in security events or potential insider threats.

- Event logs: Significant occurrences inside IOMT systems, such as software installations, system restarts, or major faults, are recorded in event logs. These logs keep a chronological record of system-level occurrences and the details that go with them. Event log analysis assists investigators in understanding system activity, identifying system vulnerabilities, and tracking the recurrence of certain events that may be important to a forensic inquiry. Event logs can reveal system misconfigurations, possible malware infestations, and anomalous system activity.

Forensic study of metadata and log files necessitates knowledge of data interpretation, log analysis tools, and IOMT system designs; however, there are many challenges as can be seen in Fig. 9.5. Investigators can develop a timeline of events, reconstruct incident scenarios, identify accountable individuals, and gather essential evidence for legal processes or incident response operations by utilizing the information contained in metadata and log files. In IOMT settings, metadata and log files contribute considerably to the whole forensic analysis process, assuring the integrity, correctness, and dependability of the investigative results. Fig. 9.5 showcases various digital forensics challenges in IOMT.

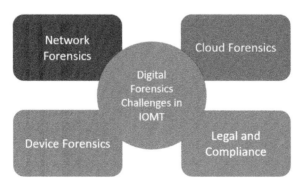

Figure 9.5 Digital forensics challenges in Internet of Medical Things (IOMT).

9.4 Forensic investigation process in Internet of Medical Things

In the context of the IOMT the forensic investigation process follows a methodical methodology to unearth digital evidence, analyze occurrences, and recreate events. It consists of various processes, including the identification and preservation of digital evidence from linked medical equipment and systems, data gathering, analysis using specialized tools and methodologies, and reporting of findings. This procedure is intended to protect the integrity and authenticity of evidence, to aid in incident response, and to give critical insights for legal proceedings or remedial efforts. Forensic investigators may efficiently negotiate the intricacies of IOMT settings and give accurate and dependable results by using this organized technique [37].

9.4.1 Step-by-step overview of the forensic investigation process specific to Internet of Medical Things environments

To provide accurate and trustworthy results, conducting a forensic investigation in the context of the IOMT requires a systematic and well-defined approach [38]. The following is a step-by-step breakdown of the forensic investigation procedure in IOMT environments:

Step 1: Planning and identification
- Define the inquiry's scope and objectives, including the event or security breach under investigation.
- Define the investigation's resources, skills, and tools.
- Create a clear strategy defining the timeframe, tasks, and responsibilities of the inquiry.

Step 2: Digital evidence preservation
- Locate and save pertinent digital evidence sources such as device logs, network traffic, sensor data, metadata, and log files.
- Use correct preservation measures to keep the evidence's integrity and validity.
- Make forensic pictures or duplicates of devices and storage media to prevent data contamination or manipulation.

Step 3: Gathering and acquiring
- Using forensically sound approaches, collect digital evidence from preserved sources.
- Collect data from linked medical devices, servers, network devices, and related storage media using specialized tools and methodologies.

- Maintain tight documentation and a chain of custody to verify the admissibility and integrity of evidence obtained.
  Step 4: Analyze and examine
- Analyze the gathered digital evidence thoroughly, including metadata, log files, sensor data, and related artifacts.
- Examine the evidence for patterns, abnormalities, or signs of malicious behavior using proper forensic analysis techniques and tools.
- Correlate various forms of data to determine the chronology of events, identify potential security breaches, and get important insights.
  Step 5: Visualization and reconstruction
- Rebuild the timeline and sequence of events that led up to the incident or security breach.
- Present the findings in a clear and succinct manner by using visualization tools such as timelines, flowcharts, or diagrams.
- Determine the incident's cause and effect, including any vulnerabilities, compromised devices, or unauthorized access points.
  Step 6: Interpretation and analysis
- Interpret the findings based on the digital evidence analysis and investigation methodologies used.
- Assess the evidence's importance in connection to the incident, security rules, and regulatory obligations.
- Draw findings and potential solutions for avoiding future threats and increasing IOMT system security.
  Step 7: Reporting and documentation
- Record the whole investigative process, including the procedures done, evidence gathered, analysis completed, and conclusions reached.
- Write a thorough forensic report summarizing the investigation, the detected incident, the evidence supporting the conclusions, and any recommendations for corrective or preventive actions.
- Ensure that the report is factual, objective, and appropriate for presentation in legal proceedings, if necessary.
  Step 8: Communication and presentation
- Inform key stakeholders, such as health-care administrators, IT workers, or legal authorities, about the results and recommendations.
- Clearly convey the technical parts of the study in a way that nontechnical folks may comprehend.
- Respond to any questions, complaints, or follow-up actions from stakeholders based on the findings of the forensic investigation.

It is critical to follow ethical and legal norms, preserve confidentiality, and protect patient privacy throughout the forensic investigation process. By following this step-by-step guide, forensic investigators will be able to successfully negotiate the specific obstacles of conducting investigations in IOMT environments, assuring the integrity, correctness, and validity of the investigative results.

9.4.2 Identification and preservation of digital evidence in a connected medical device ecosystem

In forensic investigations within a connected medical device ecosystem, the identification and preservation of digital evidence is crucial. Maintaining the integrity and admissibility of digital evidence in court processes requires accurate and safe preservation. The major stages needed in identifying and preserving digital evidence in a connected medical device ecosystem are outlined next:

1. Locate relevant digital evidence sources
   a. Begin by recognizing the many sorts of digital evidence that may exist in the linked medical device ecosystem. This covers both data kept within medical devices and data sent to and from the devices.
   b. Device logs, configuration settings, sensor data, communication records, and user activity logs are common sources of digital evidence in linked medical device ecosystems.
2. Decide on a preservation methodology
   a. Determine the preservation needs depending on the nature of the evidence and the specific instruments used.
   b. Determine if live or offline preservation is preferable. While the device is functioning, live preservation entails recording and preserving data, whereas offline preservation entails isolating and disconnecting the device from the network to avoid data tampering or contamination.
3. Utilize forensically sound techniques
   a. Use forensically sound approaches to assure digital evidence preservation. This involves the use of write-blockers or forensic imaging tools to produce exact copies of storage media without modifying the original data.
   b. Maintain a rigorous chain of custody for all evidence to ensure its validity and integrity. To guarantee adequate documentation and accountability, document each stage of the preservation process.

4. Document device specifications
   a. Compile an exhaustive list of all linked medical devices engaged in the study. Keep track of important details, including the device's make, model, serial number, firmware version, and any unique identifiers.
   b. Record the network configuration, including IP addresses, subnet information, and any other network-related information pertinent to the inquiry.
5. Capture device logs and configuration parameters
   a. Gather and save device logs, which capture significant device events and activities such as system startup, shutdown, problems, and user interactions.
   b. Record and save configuration parameters from linked medical equipment. This covers any modifications made to the default setups as well as device settings, permissions, access control lists, and access control lists.

Proper identification and preservation of digital evidence within a connected medical device ecosystem are critical for preserving the evidence's integrity and reliability. By following these processes, forensic investigators may guarantee that digital evidence is reliably collected, securely maintained, and ready for examination, allowing for a more complete and effective inquiry [39].

9.4.3 Collection and analysis of data from various Internet of Medical Things components, including wearables, medical implants, and healthcare systems

The collecting and analysis of data from many components are critical in finding evidence and comprehending the complexities of occurrences in the context of forensic investigations in the IOMT. The varied spectrum of IOMT components, including wearables, medical implants, and healthcare systems, needs specialized data gathering and processing approaches and concerns. The next sections go over the data gathering and analysis procedures for each of these components:

1. Wearables
   a. Wearable gadgets, such as smartwatches and fitness trackers, collect useful information on patient health, physical activity, and environmental variables.
   b. Data collection: By connecting wearables to a computer system and utilizing specialized software or APIs supplied by device

makers, forensic investigators can capture data from wearables. Mobile applications linked to wearables might potentially be used to extract data.

   c. Analysis: Wearable data can provide information about a user's activity, heart rate, sleep habits, and GPS position history. Forensic analysis of wearable data can aid in the establishment of timelines, the corroboration or refutation of alibis, and the provision of contextual information pertinent to the inquiry.

2. Medical implants

   a. Medical implants like pacemakers, insulin pumps, and neurostimulators produce vital data on patient health and device functionality.

   b. Data collection: Data collection from medical implants necessitates the use of specialized technologies and skills. To get data directly from implanted devices or through the linked medical systems, forensic investigators may collaborate closely with medical practitioners.

   c. Analysis: Medical implant data analysis can provide information about device programming, patient health factors, and probable abnormalities or malfunctions. Investigators can review the collected data to see if there were any abnormalities or unauthorized alterations that could be pertinent to the inquiry.

3. Health-care systems

   a. EHRs, hospital information systems, and other databases containing patient and medical data are all part of health-care systems.

   b. Data collecting: Typically, data collection from health-care systems entails accessing and retrieving pertinent information from databases. This may necessitate obtaining the necessary approvals and working with health-care providers or IT employees.

   c. Data analysis: Data from health-care systems may be analyzed to provide vital information such as patient records, medical histories, drug prescriptions, and treatment plans. Investigators can detect any unauthorized access, changes, or data breaches in health-care systems, giving crucial evidence for forensic investigation.

Forensic examination of data generated from various IOMT components necessitates the use of specialized tools, device architectural expertise, and an awareness of medical data formats and standards. To maintain the integrity and dependability of the obtained data, investigators must follow best practices [40].

9.5 Forensic techniques and tools for Internet of Medical Things

In the context of the IOMT the forensic investigation process follows a methodical methodology to unearth digital evidence, analyze occurrences, and recreate events. The digital forensic tools and techniques that are used in IOMT are shown in Fig. 9.6. It consists of various processes, including the identification and preservation of digital evidence from linked medical equipment and systems, data gathering, analysis using specialized tools and methodologies, and reporting of findings. This procedure is intended to protect the integrity and authenticity of evidence, to aid in incident response, and to give critical insights for legal proceedings or remedial efforts. Forensic investigators may efficiently negotiate the intricacies of IOMT settings and give accurate and trustworthy results by using this organized strategy [41].

9.5.1 Overview of specialized forensic techniques and tools used in Internet of Medical Things investigations

To properly analyze digital evidence and reveal key insights in the setting of the IOMT, forensic investigations require specialized methodologies and tools. Because of the unique properties of IOMT systems, such as the multitude of devices, data formats, and connectivity, forensic methodologies and tools customized to this domain are required [42]. The following

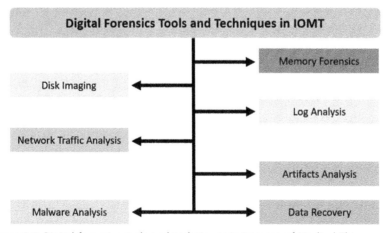

Figure 9.6 Digital forensics tools and techniques in Internet of Medical Things.

is an outline of common specialized forensic methods and instruments used in IOMT investigations:

1. Device acquisition and imaging
   a. Forensic investigators collect data from linked medical devices, wearables, and other IOMT components using specialized tools and methodologies. This entails producing forensic images or duplicates of storage media to ensure the integrity and preservation of the original data.
   b. For device collection and imaging, tools such as the Cellebrite Physical Analyzer, Magnet AXIOM, and AccessData FTK Imager are often utilized.
2. Data recovery and carving
   a. Data recovery and carving procedures are used to retrieve data from storage media when the file system has been corrupted or erased. These procedures aid in the recovery of deleted files, fragments, and data remains that may be significant to the inquiry.
   b. In IOMT investigations, tools like EnCase, Scalpel, or PhotoRec are typically utilized for data recovery and carving.
3. Network traffic analysis
   a. Understanding communication patterns, spotting potential security breaches, and analyzing data flows between linked medical devices and other systems all rely on network traffic analysis.
   b. In IOMT investigations, tools like Wireshark, NetworkMiner, and Suricata are routinely used to capture and analyze network data.
4. Metadata analysis
   a. Metadata connected with digital files and conversations can help forensic investigators. Examining timestamps, file properties, geolocation data, and other information to build timelines, correlations, and possible proof of unauthorized behavior is what metadata analysis entails.
   b. ExifTool, Metadata Analyzer, and OSForensics are often used in IOMT investigations for metadata analysis.
5. Data decryption and password recovery
   a. When confronted with encrypted data or password-protected devices, forensic investigators use procedures to decrypt data or recover passwords to obtain access to the necessary information.
   b. In IOMT investigations, tools like Elcomsoft Password Recovery, Passware Kit Forensic, and John the Ripper are routinely utilized for data decryption and password recovery.

9.5.2 Data acquisition and analysis tools for extracting evidence from diverse Internet of Medical Things devices

The IOMT refers to a diverse set of devices, each with its own set of data formats, communication protocols, and storage systems. Specialized tools are necessary during forensic investigations to efficiently retrieve evidence from various IOMT devices. These technologies help in the collection, storage, and interpretation of digital evidence [38]. The following are some examples of common data collecting and processing methods used to extract evidence from various IOMT devices:

1. Cellebrite Physical Analyzer: Cellebrite Physical Analyzer is a comprehensive forensic tool that is commonly used to acquire and analyze digital evidence from various IOMT devices. It allows data to be extracted and decoded from smartphones, tablets, wearables, and other connected devices. The software can read and understand data from backups, physical acquisitions, and logical extractions.

2. Magnet AXIOM: Magnet AXIOM is a well-known forensic platform that is used to collect and analyze digital evidence from IOMT devices. It can extract data from smartphones, wearables, cloud services, and IoT devices and supports a wide range of devices. Magnet AXIOM recovers lost or concealed data using complex parsing and carving algorithms, and its extensive search and analysis tools allow investigators to swiftly find pertinent evidence.

3. Oxygen Forensic Detective: Oxygen Forensic Detective is a powerful forensic tool for collecting and analyzing digital evidence from mobile devices, wearables, and cloud services. It allows data to be extracted and analyzed from smartphones, fitness trackers, smartwatches, and other IOMT devices. To help in the investigative process the application includes extensive analytics and reporting functions.

4. AccessData FTK (Forensic Toolkit): AccessData FTK is a popular digital forensic tool for acquiring and analyzing data from various IOMT devices. Investigators may use it to collect, process, and analyze data from smartphones, tablets, and wearables. FTK provides powerful search and indexing capabilities, enabling rapid evidence searching and correlation across many devices and data sources.

5. XRY: XRY is a popular mobile forensic tool for acquiring and analyzing data from smartphones, tablets, and other IOMT devices. It supports all devices and allows investigators to extract data such as call records, texts, emails, app data, and media assets. The sophisticated

decoding capabilities of XRY aid in the recovery of erased data and the discovery of critical evidence.

These data collecting and analysis software include a variety of features for obtaining evidence from various IOMT devices. Investigators can use these technologies to effectively collect data, retrieve lost information, parse diverse data formats, and analyze the recovered evidence. To properly manage the expanding world of IOMT devices and their related digital evidence, forensic specialists must remain up to speed with the newest versions and improvements in these technologies.

9.5.3 Considerations for ensuring the integrity and authenticity of collected evidence

It is critical to ensure the integrity and validity of acquired evidence while conducting forensic investigations in the IOMT environment. Maintaining the integrity and validity of digital evidence is critical for its admissibility in court proceedings and the investigation's credibility. To preserve the integrity and validity of acquired evidence, the following factors should be considered:

1. Chain of custody: Create a documented chain of custody for all evidence gathered. This involves keeping track of the date, time, place, people involved, and any modifications or transfers of custody. Maintaining a secure chain of custody protects the evidence from tampering or unauthorized access during the inquiry.

2. Proper documentation: Document all actions and procedures taken throughout evidence collecting and analysis. This includes documenting the tools used, the methods employed, and any changes made to the original data. Detailed documentation increases transparency and aids in the verification of the evidence's correctness and trustworthiness.

3. Write-blocking and read-only techniques: When obtaining data from IOMT devices, use write-blocking and read-only approaches. Write-blocking guarantees that no changes to the original data are performed throughout the acquisition process, preserving its integrity. Read-only procedures prohibit unintentional or purposeful changes to evidence during analysis.

4. Hashing and digital signatures: To produce unique hash values for collected data, use cryptographic hashing techniques such as MD5 or SHA-256. Hashing protects data integrity by ensuring that the

evidence obtained stays intact. Furthermore, digital signatures may be used to verify the origins and integrity of the evidence.

5. Time stamping: Use reliable time-stamping devices to determine when evidence was gathered or amended. Time stamping establishes the legitimacy and dependability of the evidence by providing a reference point for chronological order.

By following these guidelines, forensic investigators can assure the integrity and validity of evidence acquired in IOMT investigations. The preservation of evidence's integrity and authenticity improves its probative value, increases the credibility of the inquiry, and secures its admission in judicial processes.

9.6 Future directions and research challenges

Forensic investigation in the IOMT offers fascinating potential as well as research problems. Advances in AI and ML for more efficient analysis, the creation of standardized frameworks and protocols, enhanced incident response tactics, and user-centered design approaches are among the future objectives. However, difficulties such as data complexity, privacy preservation, evidence trust and integrity, standardization, and human aspect consideration must be addressed. Further study and collaboration are required to overcome these obstacles and develop IOMT forensics, hence improving patient safety and health-care security.

9.6.1 Exploration of emerging trends and future directions in the field of forensic investigation in Internet of Medical Things

The discipline of forensic inquiry in the IOMT is constantly changing, driven by technological improvements, new problems, and the increasing complexity of linked medical equipment [43]. It is critical to investigate current trends and future directions in this industry to remain ahead of the curve and effectively address the changing situation. Some significant topics of interest and prospective future directions in the field of IOMT forensic investigation are highlighted next:

1. AL and ML: For IOMT forensic investigations the combination of AI and ML approaches offers considerable promise. AI and ML algorithms may be trained to analyze enormous amounts of data, discover trends, detect abnormalities, and aid in the detection of possible security breaches or criminal activity. Future research may concentrate on the development of powerful AI-based tools tailored particularly for forensic investigation in IOMT contexts.

2. Blockchain technology: Blockchain technology enables data storage and administration that is decentralized, secure, and tamper-evident. Its possible use in IOMT forensic investigations is to ensure the integrity, traceability, and validity of digital evidence acquired. In IOMT investigations, research may look at the usage of blockchain-based systems for secure evidence retention, chain-of-custody management, and data integrity verification.

3. Privacy-preserving techniques: As privacy issues remain an important part of IOMT forensic investigations, developing privacy-preserving approaches becomes necessary. Future study might look into advanced cryptographic approaches like homomorphic encryption or safe multiparty computing to enable forensic analysis while protecting sensitive patient data privacy.

4. Forensic readiness frameworks: The creation of forensic preparedness frameworks tailored to IOMT settings can assist organizations in proactive preparation for forensic investigations. Guidelines for data collection, storage, recording, and incident response processes may be included in these frameworks to assist efficient and successful forensic analysis. Efforts in the future may seek to standardize and develop best practices for forensic preparedness in IOMT ecosystems.

5. Integration of IoT security and forensics: As the IOMT ecosystem grows more linked, it is critical to include IoT security and forensic capabilities. Future study might look toward the creation of unified frameworks that integrate security measures with built-in forensic capabilities. This integration would allow for the identification of security incidents in real time, fast forensic investigation, and simplified incident response [44,45].

9.6.2 Research challenges and opportunities for advancements in Internet of Medical Things forensics

Forensic investigation in the IOMT poses a number of research problems as well as potential for improvement. Addressing these problems and seizing opportunities will help to build more effective and robust forensic approaches in the IOMT context. The following are some of the most important research difficulties and possibilities in the field of IOMT forensics:

1. Complexity of IOMT ecosystems
   a. Challenge: The linked structure of IOMT ecosystems complicates forensic investigations. For successful analysis the varied range of devices, communication protocols, and data formats necessitates the use of specialized tools and procedures.

b. Opportunity: Research might be focused on building comprehensive frameworks and approaches to meet the particular difficulties offered by IOMT ecosystems. This comprises methods for identifying devices, correlating data, and recreating the digital evidence chain across numerous networked devices.

2. Data heterogeneity and volume

a. Challenge: Massive amounts of heterogeneous data are generated by IOMT from numerous sensors, wearables, and health-care systems. Analyzing and collecting useful forensic evidence from such large and varied databases are difficult.

b. Opportunity: Advanced data analytics approaches, such as data mining, ML, and pattern recognition, can be investigated to enable efficient and automated analysis of large-scale IOMT data. This involves creating algorithms to detect abnormalities, detect suspicious trends, and extract useful insights from disparate data sources.

3. Privacy-preserving techniques

a. Challenge: Patient privacy must be protected when performing forensic investigations in IOMT. Balancing the necessity for forensic investigation with the safeguarding of sensitive patient data poses distinct issues.

b. Opportunity: The focus of research might be on creating privacy-preserving strategies that allow for effective forensic analysis while reducing the danger of privacy breaches. This involves investigating cryptographic technologies, secure computing techniques, and anonymization ways to safeguard patient data during the inquiry.

4. Trust and integrity of digital evidence

a. Challenge: A significant difficulty is ensuring the trustworthiness and integrity of digital evidence acquired via IOMT devices. Adversarial attacks, manipulation, or unintended adjustments might jeopardize the evidence's credibility and admissibility.

b. Opportunity: Techniques for improving the confidence and integrity of digital evidence in IOMT forensics can be researched. This involves looking at digital signatures, blockchain technology, and tamper-evident systems to verify the validity and integrity of evidence collected during the investigation process [42].

5. Standardization and interoperability

a. Challenge: The absence of standardized forensic methodology, data formats, and compatibility across IOMT devices and forensic software makes it difficult to conduct seamless investigations.

   b. Opportunity: The development of standards and methods for IOMT forensic investigations might be the topic of research. Defining common data formats, communication protocols, and interoperability standards to promote effective data interchange, analysis, and cooperation among many parties involved in forensic investigations is part of this.
6. Incident response and recovery
   a. Challenge: In IOMT forensic investigations, rapid incident reaction and successful recovery procedures are critical. It is a huge task to develop efficient procedures for detecting, responding to, and recovering from security incidents.
   b. Opportunity: The research can concentrate on establishing incident response mechanisms tailored to IOMT settings. This comprises real-time monitoring tools, early detection of security issues, effective incident response processes, and effective recovery mechanisms to reduce the impact of security breaches on patient safety and data integrity.

9.6.3 Integration of artificial intelligence and machine learning in forensic analysis of Internet of Medical Things data

In the context of the IOMT the area of forensic analysis is increasingly harnessing the potential of AI and ML techniques to improve the analysis and interpretation of massive volumes of complicated IOMT data. The use of AI and ML in forensic analysis provides tremendous prospects for increasing the efficiency, accuracy, and depth of IOMT investigations [38]. This section investigates the use of AI and ML in the forensic analysis of IOMT data, as well as the possible benefits and obstacles.
1. Data analysis and pattern recognition: Large amounts of IOMT data, like sensor readings, patient records, and device logs, may be analyzed using AI and ML approaches to detect trends, anomalies, and possible evidence. AI can automatically discover suspicious actions, correlations, and deviations from typical behavior using powerful data mining and pattern recognition algorithms, enabling forensic investigators in spotting probable security breaches or unauthorized access.
2. Predictive analytics and decision support: AI and ML algorithms may be trained on past IOMT data to create predictive models capable of anticipating possible security attacks, identifying vulnerabilities, and forecasting aberrant device behavior. These predictive skills can enable

proactive incident prevention, incident response, and decision-making during forensic investigations.

3. Anomaly detection and intrusion detection: AI and ML approaches may be used to create baseline behavior models for IOMT devices and networks. These models can detect irregularities and possible intrusions in the IOMT ecosystem by continually monitoring and analyzing real-time data. This enables forensic investigators to immediately spot suspicious activity, malicious behavior, or unauthorized access attempts, allowing for faster incident response and forensic investigation.

4. Data fusion and correlation: AI and ML can combine and correlate data from many IOMT devices and sources, allowing for a holistic picture of the ecosystem. AI algorithms can extract meaningful insights, identify relationships, and establish temporal correlations between events by fusing data from wearable devices, medical implants, healthcare systems, and other relevant sources, assisting in the reconstruction of incidents and establishing the digital evidence chain.

5. Natural language processing and text analysis: Natural language processing approaches, for example, can aid in the analysis of unstructured data sources such as medical records, clinical notes, or communication logs. NLP can assist forensic investigators in collecting essential evidence, identifying significant players or entities, and comprehending contextual information pertinent to the inquiry by extracting relevant information, identifying keywords, and categorizing text [46].

6. Image and video analysis: Forensic analysis may be performed using AI and ML algorithms that analyze visual data such as photos or video footage acquired by IOMT devices or surveillance systems. Computer vision techniques can recognize objects, faces, gestures, or particular events, which can help with the identification of pertinent evidence, the identification of persons, or the reconstruction of timelines during investigations. It takes a large amount of time and space to process high-resolution medical images. To address these limits, numerous image retargeting techniques have been developed to minimize the size of medical photos while maintaining visual quality [46—48].

Finally, using AI and ML into forensic analysis of IOMT data has enormous prospects for enhancing investigative efficiency and efficacy. AI can help forensic investigators extract useful insights, spot anomalies, and locate possible evidence by utilizing sophisticated analytics, pattern recognition, and automation. However, concerns connected to data quality, openness, ethics, and security must be addressed with caution. Continued study, cooperation, and the creation of ethical frameworks will influence the appropriate and successful application of AI and ML in the field of IOMT forensics, contributing to enhanced patient safety and health-care security (Tables 9.1 and 9.2).

9.7 Conclusion

Forensic investigation in the IOMT plays a vital role in addressing cybersecurity challenges and ensuring the integrity and security of patient data. This chapter has provided an in-depth exploration of forensics in IOMT, highlighting the unique considerations and challenges involved in conducting forensic analysis of connected medical devices. The preservation of digital evidence has been emphasized as a crucial aspect of IOMT investigations, enabling the reconstruction of incidents and maintaining the integrity of

Table 9.1 Summary of challenges and opportunities.

Challenges	Opportunities
Complexity of IOMT ecosystems	• Building comprehensive frameworks and approaches to analyze IOMT ecosystems.
Data heterogeneity and volume	• Investigating advanced data analytics approaches such as data mining, machine learning, and pattern recognition for efficient analysis of large-scale IOMT data.
Privacy-preserving techniques	• Creating privacy-preserving strategies that allow for effective forensic analysis while reducing the risk of privacy breaches.
Trust and integrity of digital evidence	• Researching techniques to improve the confidence and integrity of digital evidence in IOMT forensics, such as digital signatures, blockchain technology, and tamper-evident systems.
Standardization and interoperability	• Developing standards and methods for IOMT forensic investigations, including common data formats, communication protocols, and interoperability standards.
Incident response and recovery	• Establishing incident response mechanisms tailored to IOMT settings, including real-time monitoring tools, early detection of security issues, and effective incident response and recovery processes.

Table 9.2 Use of artificial intelligence (AI) and machine learning (ML) in the forensic analysis.

Use of AI and ML	Relevant information
Data analysis and pattern recognition	• AI and ML can analyze large amounts of IOMT data to detect trends, anomalies, and possible evidence. • Help identify security breaches or unauthorized access.
Predictive analytics and decision support	• Trains AI and ML algorithms on past IOMT data to create predictive models. • Anticipates security attacks, identifies vulnerabilities, and forecasts aberrant device behavior. • Enables proactive incident prevention.
Anomaly detection and intrusion detection	• AI and ML create baseline behavior models for IOMT devices and networks. • Continuously monitor real-time data for irregularities and intrusions. • Aid in faster incident response and forensic investigation.
Data fusion and correlation	• AI and ML combine and correlate data from multiple IOMT devices and sources. • Provides a holistic view of the ecosystem. • Assists in reconstructing incidents and establishing the digital evidence chain.
Natural language processing and text analysis	• NLP aids in analyzing unstructured data sources like medical records and communication logs. • Extracts relevant information and identifies keywords. • Helps comprehend contextual information.
Image and video analysis	• AI and ML algorithms analyze visual data from IOMT devices and surveillance systems. • Recognizes objects, faces, gestures, or events. • Assists in identifying evidence and reconstructing timelines.

findings. The chapter discussed various types of digital evidence encountered in IOMT systems, ranging from sensor data and health-related information to metadata and log files. The forensic investigation process specific to IOMT environments was presented, providing a step-by-step overview of the identification, preservation, collection, and analysis of digital evidence from different IOMT components. The importance of specialized forensic techniques and tools for IOMT investigations was underscored, with a focus on data acquisition and analysis tools for diverse IOMT devices. Looking toward the future, the integration of AI and ML in forensic analysis of IOMT data holds immense potential. These technologies can enhance investigative capabilities, aid in anomaly detection, and facilitate predictive analytics for proactive incident prevention. However, research challenges and opportunities in the field of IOMT forensics need to be addressed, paving the way for advancements in security, privacy, and digital evidence preservation. In conclusion, forensic investigation in IOMT is crucial for safeguarding the integrity and security of connected medical devices and the sensitive data they handle. Continued research and development in this field are essential to stay ahead of emerging threats and to ensure the trustworthiness of IOMT systems in the health-care domain. This chapter delves at forensic investigation in the context of the IOMT. Throughout the chapter, it addressed numerous crucial themes. To begin the chapter looked at the concept and scope of forensic inquiry in IOMT, emphasizing the need of preserving the integrity and security of IOMT systems. It also looked at the difficulties and special considerations that come with doing forensic analysis in IOMT contexts, such as data heterogeneity and privacy issues. The chapter also emphasized the need of digital evidence preservation in IOMT environments and gave a step-by-step description of the forensic investigation procedure in IOMT. In addition, the chapter examined specialized forensic techniques and technologies utilized in IOMT investigations, as well as the use of AI and ML in IOMT forensic analysis. The chapter also discussed the legal, ethical, and privacy concerns of performing forensic investigations in IOMT, with an emphasis on patient data security and regulatory framework compliance. Furthermore, the chapter investigated new trends, future directions, research obstacles, and prospects for breakthroughs in IOMT forensics. Finally, we emphasized the necessity of forensic investigation in preserving trust and security in IOMT systems, as well as the need for more study and collaboration in the sector. Overall, this chapter gives a thorough review of forensic investigation in IOMT, emphasizing its importance, problems, breakthroughs, and the need for continuing study and collaboration.

References

[1] R. Rajkumar, I. Lee, L. Sha, J. Stankovic, Cyber-physical systems: the next computing revolution, in: Proceedings of the 47th Design Automation Conference, 2010, pp. 731–736.

[2] P.A. Laplante, N. Laplante, The internet of things in healthcare: potential applications and challenges, It Professional 18 (3) (2016) 2–4.

[3] P.K. Sadhu, V.P. Yanambaka, A. Abdelgawad, Internet of Things: security and solutions survey, Sensors 22 (19) (2022) 7433.

[4] P. Tiwari, V. Garg, R. Agrawal, Changing world: smart homes review and future, Smart IoT for Research and Industry (2022) 145–160.

[5] R. Dwivedi, D. Mehrotra, S. Chandra, Potential of Internet of Medical Things (IoMT) applications in building a smart healthcare system: a systematic review, Journal of Oral Biology and Craniofacial Research 12 (2) (2022) 302–318.

[6] S. Krishnamoorthy, A. Dua, S. Gupta, Role of emerging technologies in future IoT-driven Healthcare 4.0 technologies: a survey, current challenges and future directions, Journal of Ambient Intelligence and Humanized Computing 14 (1) (2023) 361–407.

[7] R. Mahmud, R. Kotagiri, R. Buyya, Fog computing: a taxonomy, survey and future directions, Internet of Everything: Algorithms, Methodologies, Technologies and Perspectives, Springer, 2018, pp. 103–130.

[8] J. Sametinger, J. Rozenblit, R. Lysecky, P. Ott, Security challenges for medical devices, Communications of the ACM 58 (4) (2015) 74–82.

[9] A. Ghubaish, T. Salman, M. Zolanvari, D. Unal, A. Al-Ali, R. Jain, Recent advances in the internet-of-medical-things (IoMT) systems security, IEEE Internet of Things Journal 8 (11) (2020) 8707–8718.

[10] K.K. Karmakar, V. Varadharajan, U. Tupakula, S. Nepal, C. Thapa, Towards a security enhanced virtualised network infrastructure for Internet of Medical Things (IoMT), in: 2020 Sixth IEEE Conference on Network Softwarization (NetSoft), IEEE, 2020, pp. 257–261.

[11] F. Rahman, N. Shabana, Wireless sensor network based personal health monitoring system, WSEAS Transactions on Systems 5 (5) (2006) 966–972.

[12] E. McMahon, R. Williams, M. El, S. Samtani, M. Patton, H. Chen, Assessing medical device vulnerabilities on the Internet of Things, in: 2017 IEEE International Conference on Intelligence and Security Informatics (ISI), IEEE, 2017, pp. 176–178.

[13] P. Galetsi, K. Katsaliaki, S. Kumar, Values, challenges and future directions of big data analytics in healthcare: a systematic review, Social Science & Medicine 241 (2019) 112533.

[14] J.D. Halamka, M. Tripathi, The HITECH era in retrospect, The New England Journal of Medicine 377 (10) (2017) 907–909.

[15] J.-P.A. Yaacoub, M. Noura, H.N. Noura, O. Salman, E. Yaacoub, R. Couturier, et al., Securing internet of medical things systems: limitations, issues and recommendations, Future Generation Computer Systems 105 (2020) 581–606.

[16] H. Suo, J. Wan, C. Zou, J. Liu, Security in the internet of things: a review, in: 2012 International Conference on Computer Science and Electronics Engineering, vol. 3, IEEE, 2012, pp. 648–651.

[17] I. Yaqoob, I.A.T. Hashem, A. Ahmed, S.M.A. Kazmi, C.S. Hong, Internet of things forensics: recent advances, taxonomy, requirements, and open challenges, Future Generation Computer Systems 92 (2019) 265–275.

[18] S. Fugkeaw, P. Sanchol, Enabling efficient personally identifiable information detection with automatic consent discovery, ECTI Transactions on Computer and Information Technology (ECTI-CIT) 17 (2) (2023) 245–254.

[19] M. Al Ameen, J. Liu, K. Kwak, Security and privacy issues in wireless sensor networks for healthcare applications, Journal of Medical Systems 36 (2012) 93–101.

[20] J. Hou, Y. Li, J. Yu, W. Shi, A survey on digital forensics in Internet of Things, IEEE Internet of Things Journal 7 (1) (2019) 1–15.

[21] A.M.K. Marshall, Digital Forensics: Digital Evidence in Criminal Investigations, John Wiley & Sons, 2009.

[22] B. Carrier, File System Forensic Analysis, Addison-Wesley Professional, 2005.

[23] T. Jabeen, H. Ashraf, A. Ullah, A survey on healthcare data security in wireless body area networks, Journal of Ambient Intelligence and Humanized Computing 12 (2021) 9841−9854.

[24] H. Arshad, A.B. Jantan, O.I. Abiodun, Digital forensics: review of issues in scientific validation of digital evidence, Journal of Information Processing Systems 14 (2) (2018) 346−376.

[25] A. Arora, A. Nandkumar, R. Telang, Does information security attack frequency increase with vulnerability disclosure? An empirical analysis, Information Systems Frontiers 8 (2006) 350−362.

[26] S. Zawoad, R. Hasan, Cloud forensics: a meta-study of challenges, approaches, and open problems. arXiv preprint arXiv:1302.6312, 2013.

[27] O.I. Abiodun, E.O. Abiodun, M. Alawida, R.S. Alkhawaldeh, H. Arshad, A review on the security of the internet of things: challenges and solutions, Wireless Personal Communications 119 (2021) 2603−2637.

[28] V. Mohammadi, A.M. Rahmani, A.M. Darwesh, A. Sahafi, Trust-based recommendation systems in Internet of Things: a systematic literature review, Human-Centric Computing and Information Sciences 9 (1) (2019) 1−61.

[29] H. Kim, D. Kim, M. Kwon, H. Han, Y. Jang, D. Han, et al., Breaking and fixing volte: exploiting hidden data channels and mis-implementations, in: Proceedings of the 22nd ACM SIGSAC Conference on Computer and Communications Security, 2015, pp. 328−339.

[30] M. Conti, A. Dehghantanha, K. Franke, S. Watson, Internet of Things security and forensics: challenges and opportunities, Future Generation Computer Systems 78 (2018) 544−546.

[31] G. Thamilarasu, A. Odesile, A. Hoang, An intrusion detection system for internet of medical things, IEEE Access 8 (2020) 181560−181576.

[32] A. Kosba, A. Miller, E. Shi, Z. Wen, C. Papamanthou, Hawk: the blockchain model of cryptography and privacy-preserving smart contracts, in: 2016 IEEE Symposium on Security and Privacy (SP), IEEE, 2016, pp. 839−858.

[33] Y. Yang, L. Wu, G. Yin, L. Li, H. Zhao, A survey on security and privacy issues in Internet-of-Things, IEEE Internet of Things Journal 4 (5) (2017) 1250−1258.

[34] O. AlShorman, B. AlShorman, M. Al-khassaweneh, F. Alkahtani, A review of internet of medical things (IoMT)-based remote health monitoring through wearable sensors: a case study for diabetic patients, Indonesian Journal of Electrical Engineering and Computer Science 20 (1) (2020) 414−422.

[35] S. Jain, M. Nehra, R. Kumar, N. Dilbaghi, T.Y. Hu, S. Kumar, et al., Internet of medical things (IoMT)-integrated biosensors for point-of-care testing of infectious diseases, Biosensors and Bioelectronics 179 (2021) 113074.

[36] J. Wang, K.L. Ming, C. Wang, T. Ming-Lang, The evolution of the Internet of Things (IoT) over the past 20 years, Computers & Industrial Engineering 155 (2021) 107174.

[37] F. Al-Turjman, M.H. Nawaz, U.D. Ulusar, Intelligence in the Internet of Medical Things era: a systematic review of current and future trends, Computer Communications 150 (2020) 644−660.

[38] H. Jahankhani, J. Ibarra, Digital forensic investigation for the Internet of Medical Things (IoMT), Journal of Forensic Legal & Investigative Sciences 5 (2) (2019) 029.

[39] A. Mishra, P. Bagade, Digital forensics for Medical Internet of Things, in: 2022 IEEE Globecom Workshops (GC Wkshps), IEEE, 2022, pp. 1074−1079.

[40] S.K. Polu, S.K. Polu, IoMT based smart health care monitoring system, International Journal for Innovative Research in Science & Technology 5 (11) (2019) 58−64.

[41] A.R. Javed, W. Ahmed, M. Alazab, Z. Jalil, K. Kifayat, T.R. Gadekallu, A comprehensive survey on computer forensics: state-of-the-art, tools, techniques, challenges, and future directions, IEEE Access 10 (2022) 11065−11089.

[42] J.-P.A. Yaacoub, N.N. Hassan, O. Salman, A. Chehab, Advanced digital forensics and anti-digital forensics for IoT systems: Techniques, limitations and recommendations, Internet of Things 19 (2022) 100544.

[43] A. Garg, A.K. Singh, Internet of Things (IoT): security, cybercrimes, and digital forensics, in: K. Kaushik, S. Dahiya, A. Bhardwaj, Y. Maleh (Eds.), Internet of Things and Cyber Physical Systems: Security and Forensics, CRC Press, 2022, pp. 23–50.

[44] A. Garg, A.K. Singh, Applications of Internet of Things (IoT) in green computing, Intelligence of Things: AI-IoT Based Critical-Applications and Innovations, Springer, 2021, pp. 1–34.

[45] A.K. Singh, A. Nayyar, A. Garg, A secure elliptic curve based anonymous authentication and key establishment mechanism for IoT and cloud, Multimedia Tools and Applications 82 (2022) 22525–22576.

[46] A. Garg, A. Negi, P. Jindal, Structure preservation of image using an efficient content-aware image retargeting technique, Signal, Image and Video Processing 15 (2021) 185–193.

[47] A. Garg, A. Negi, Structure preservation in content-aware image retargeting using multi-operator, IET Image Processing 14 (13) (2020) 2965–2975.

[48] A. Garg, A. Nayyar, A.K. Singh, Improved seam carving for structure preservation using efficient energy function, Multimedia Tools and Applications 81 (9) (2022) 12883–12924.

CHAPTER 10

Artificial intelligence driven cybersecurity in digital healthcare frameworks

Pragyan Das, Ishika Gupta and Sushruta Mishra
Kalinga Institute of Industrial Technology, Deemed to be University, Bhubaneswar, Odisha, India

10.1 Introduction

The domain of information technology dedicated to ensuring the security of health-care infrastructures is termed health-care cybersecurity. This category encompasses an array of systems such as electronic health records (EHRs), wearable fitness devices, medical equipment inventory, and software designed to facilitate health-care administration and delivery. Despite the notable capability of artificial intelligence (AI) to execute healthcare-related functions on par with or surpassing human performance, the adoption of extensive automation for health-care practitioner roles is anticipated to be restrained by various implementation considerations, thus deferring widespread implementation for a substantial duration [1]. The treatment of cancer diagnosis was enhanced by the usage of AI [2]. The primary aim underlying health-care cybersecurity revolves around fortifying systems against unauthorized infiltrations, thereby thwarting unapproved entry, utilization, and exposure of patient information. The overarching objective is to ensure the unwavering accessibility, confidentiality, and unaltered state of critical medical data. The compromise of such data through hacking could potentially jeopardize patient well-being, given the substantial value it holds for both nation-state entities and cybercriminals, encompassing monetary gains and intelligence insights. This renders health-care institutions acutely susceptible to cyber assaults, making them prime targets. Among the data types being targeted, one finds protected health information (PHI), financial records encompassing credit card and bank particulars, as well as personally identifiable information (PII) like Social Security numbers.

Securing Next-Generation Connected Healthcare Systems
DOI: https://doi.org/10.1016/B978-0-443-13951-2.00002-7

Given the extensive repository of valuable information at their disposal, health-care organizations become notably susceptible to cyberattacks, rendering them prime targets for both nation-state entities and cybercriminals [3]. The allure of this information stems from its multifaceted worth in terms of financial gains and strategic insights. The data in focus span a comprehensive spectrum, encompassing elements such as PHI, financial records entailing bank account and credit card specifics, PII such as Social Security numbers, and intellectual assets linked to advancements in medical research and innovation. A sample demonstration of the usage of AI in smart clinical system is shown in Fig. 10.1.

In reality, pilfered health-care information can command prices up to 10-fold higher on the dark web compared to purloined credit card data. Considering an average expense of $408 for addressing each pilfered health record, in contrast to $148 for every nonhealth-related record, the financial outlay incurred in rectifying breaches within the health-care sector exceeds that of other industries by over threefold. Regrettably, further unfavorable developments confront health-care entities. Hospitals are prime targets for hackers since they house hundreds or even thousands of patients, which is a major source of concern for hospital managers about health-care cybersecurity.

The realm of hospital cybersecurity is notably affected by ransomware attacks, a fact exemplified by the incident in 2018 that targeted Hancock Regional Hospital in Greenfield, Indiana. During this breach, hackers successfully infiltrated the backup system's data repository, subsequently inflicting irreversible harm to various files, including electronic health records (EHRs). Thankfully, patient care remained uninterrupted, as the hospital managed to sustain its operations despite the network shutdown

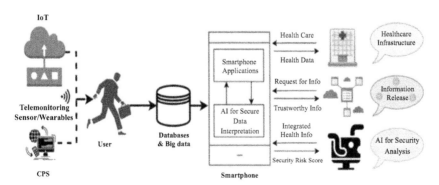

Figure 10.1 AI for smart m-HealthCare.

enacted by the IT department. However, the aftermath of the attack induced a fiscal burden on the institution, necessitating the disbursement of a ransom amounting to $55,000, transacted in four units of the digital currency Bitcoin, to effectuate the restitution of its compromised data. A myriad of manifestations characterize cyberattacks, ranging from the coercive tactics of ransomware to the illicit acquisition of individualized information. The magnitude of the establishment in question holds the potential to modulate the extent of ramifications ensuing from such assaults. Within the health-care domain, four predominant challenges garner attention: safeguarding patient confidentiality, addressing vulnerabilities inherent in legacy systems specific to healthcare, grappling with information technology challenges unique to the health-care sector, and contending with instances of security breaches that transpire within health-care environments. These concerns pervade the entirety of the industry landscape.

The main contributions of this chapter are as follows:

- The chapter is focused on the use of AI approaches in promptly recognizing and managing cyber risks in health-care environments.
- The work examines a system integrating both intrusion detection and threat intelligence and evaluation of a real-time detection system using an ensemble classifier based on network traffic features.
- A detailed comparison of two distinct AI-driven cybersecurity systems specifically designed for health-care settings is performed.
- Evaluation of attack detection accuracy, F1-score, communication cost, ROC, efficiency ratio, prediction ratio, precision, and false alarm to assess the behavior of the proposed techniques is carried out.
- A contrasting analysis between the proposed methods against conventional mechanisms to establish the superior efficiency of the newer models is done.
- The future trajectory of AI-driven cybersecurity in healthcare, emphasizing advanced machine learning techniques and responsiveness to threats, is highlighted.

10.2 Machine intelligence and cybersecurity in healthcare

AI technologies can enhance cybersecurity in healthcare by analyzing large datasets, identifying patterns and anomalies, and automating security processes. Here are some specific examples:

Threat detection and prevention: AI possesses the capability to discern and avert cyber threats through its capacity to scrutinize network traffic, discerning anomalies that might signify an impending security breach and warrant intervention. By analyzing historical data and identifying patterns, AI can also recognize new and emerging threats and help health-care organizations respond quickly to prevent data breaches.

Anomaly detection: AI can detect anomalies in data by analyzing user behavior and network traffic. For example, AI can identify unusual access attempts or data transfers that may indicate a security breach. By detecting anomalies in data, AI can help health-care organizations identify potential security threats before they can cause harm. Predictive analysis: AI can analyze historical data to predict future cyber threats and vulnerabilities. By identifying patterns in historical data, AI has the potential to assist health-care entities in proactively readying themselves for potential security threats and implementing preemptive measures to thwart such incidents.

Incident response: AI can automate incident response processes by prioritizing security incidents and automating response processes, such as patch management and vulnerability scanning. By automating incident response, AI can help health-care organizations respond quickly to security incidents and prevent data breaches.

Fraud detection: AI can detect fraud and abuse in health-care data by analyzing patient data and claims data. By identifying unusual patterns and activity, AI can help health-care organizations prevent financial losses and protect patient data.

Machine learning algorithms can undergo training to discern atypical access endeavors, data transfers, and assorted activities that could potentially signal a cyber assault. Through the identification of these irregularities, machine learning stands to aid health-care establishments in promptly and adeptly pinpointing security breaches, thereby expediting and enhancing the efficacy of their response protocols.

In cybersecurity, deep learning can be used to analyze images, text, and other unstructured data to detect and prevent cyberattacks. For example, deep learning algorithms could be trained to examine network traffic and identify malware or other types of malicious software. Deep learning can also be used to identify vulnerabilities in software and systems and help health-care organizations address these vulnerabilities before they can be exploited by cyberattackers. Fig. 10.2 shows a sensory module—enabled smart health channel.

In cybersecurity, reinforcement learning can be used to train AI agents to detect and respond to security threats. For example, reinforcement

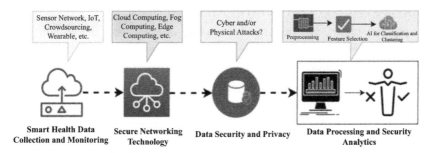

Figure 10.2 A smart health pipeline.

learning can be used to train AI agents to recognize and respond to cyber-attacks in real time. By using reinforcement learning, health-care organizations can improve their ability to respond to cyberattacks quickly and effectively. As the health-care industry continues to digitize and adopt new technologies, the need for effective cybersecurity measures becomes increasingly important. AI-driven cybersecurity solutions have the potential to strengthen the security of health-care systems and shield patient data from cyber threats. However, there are also moral and legal questions raised by the use of AI in health-care cybersecurity, such as privacy, bias, and accountability. Ongoing research and collaboration between stakeholders are necessary to ensure that AI-driven cybersecurity solutions are safe, effective, and ethical.

10.3 Literature review and background works

Several studies have proposed the use of AI in health-care cybersecurity. The literature review on intelligent techniques and machine learning applications in protecting and improving health-care systems draws from various academic sources. Al-Shaher et al.'s [3] study looks into the development of a health-care system utilizing intelligent techniques, emphasizing the role of machine learning algorithms in enhancing data correctness and reducing data redundancy. In a related development, Marwan et al. [4] conduct research on the application of machine learning in enhancing cloud-based health-care security. The introduction of neural networks is also a huge discovery. They highlight the potential of machine learning in incorporating a safety automation system with an aim to protect sensitive patient data stored in the cloud. Furthermore, studies of Kour [5] and Qu et al. [6] delve into the energy efficiency of routing algorithms for wireless body area networks (WBAN). These networks are essential in connecting various wearable and implantable medical devices, and the highlighted research shows that energy efficient algorithms are critical for the reliable and secure transmission of health data. On the aspect of cyberattack detection in healthcare, AlZubi et al. [7], Aygun and Yavuz [8], and Kumar et al. [9] have conducted studies demonstrating the efficacy of machine learning in detecting and thwarting cybersecurity threats. Advanced anomaly detection models are proposed to protect critical health data and devices against potential breaches. Looking at research by Christov and Bortolan [10], it uses pattern recognition and neural networks to classify premature ventricular contractions, providing a clear example of machine learning's applications in cardiology and diagnostic medicine. In the same vein, research projects by Hinneburg et al. [11] and Kreuseler et al. offer insight into visual mining and data visualization which can be instrumental in facilitating visual complexity reduction and pattern exploration in the vast data sphere. Finally, studies of Meng et al. [12] and Ardito et al. [13] delve into trust management involving Bayesian-based analysis to counter insider attacks, particularly in the context of software-defined networks in healthcare. Overall, this literature review suggests an active academic interest in combining machine learning, data visualization, and intelligent techniques to bolster health-care data security and enhance the efficiency and effectiveness of health-care systems.

10.4 Challenging issues of cybersecurity in healthcare

Generic privacy and security issues in cyber secure systems are presented in Fig. 10.3. Phishing, malware, and hacked IoT devices are quite common risks prevalent in today's era. Common issues are discussed next.

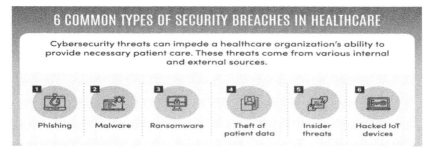

Figure 10.3 Common security breaches in healthcare.

- The development and widespread use of internet-connected medical devices have occurred without the implementation of adequate privacy and security protections. There are no universally agreed-upon industry standards for privacy and security in the development of IoT medical devices. There is currently no comprehensive regulatory framework for IoT medical devices. This means that manufacturers are not required to adhere to specific privacy and security standards, which can lead to inconsistent practices across the industry.
- The proliferation of unregulated mobile applications that utilize PHI and PII without implementing adequate security measures to ensure their protection. It is relatively easy for developers to create mobile health apps without any formal training in healthcare or privacy/security regulations. This can lead to the creation of apps that do not adequately protect user data.
- Many health-care personnel are undertrained on cybersecurity. The lack of emphasis on cybersecurity training, rapidly evolving threat landscape, limited resources, complexity of healthcare IT systems, and high turnover rates all contribute to the undertraining of health-care personnel.
- Operational requirements frequently prioritize rapidity and the dissemination of information over the imperative of ensuring information security. The pressure to deliver timely care, need for real-time information, limited resources, culture of collaboration, and misconceptions about security all contribute to the prioritization of information sharing and speed.
- Large-scale data portability demands are a result of business and regulatory requirements. This happens because the need for large-scale data portability is driven by a range of business and compliance requirements, including mergers and acquisitions, regulatory requirements, business continuity, cloud adoption, and data analytics.
- The process of digitalizing stand-alone technologies and assimilating them into broader systems gives rise to a spectrum of cybersecurity challenges, including, but not limited to, dependencies linked to interoperability, apprehensions regarding network segmentation, and various other complexities. These challenges are compounded by the intricate nature of IT ecosystems, the absence of universally standardized protocols, susceptibilities inherent in legacy systems, the involvement of third-party entities, and the expanded surface available for potential attacks.

- Obsolete legacy systems, which are no longer subject to support from their manufacturers, become incapable of integrating the latest security updates. Consequently, these systems introduce enduring vulnerabilities into the network infrastructure of organizations. Legacy systems introduce permanent vulnerabilities into organizations' networks because of outdated hardware and software, lack of resources, interdependence with other systems, compliance requirements, and lack of awareness.
- A significant proportion of organizations' constrained IT budgets is allocated toward procuring, implementing, and integrating technical solutions, leaving minimal resources for the vital tasks of fortifying and sustaining network capabilities. Additionally, the organizations sometimes engage in ad hoc activities without possessing an internal focal point responsible for ensuring security.
- Credentials serve as a repetitive and ongoing means for infiltrating targeted systems, affording malicious actors numerous pathways through which they can perpetrate harm. Once a malicious actor acquires valid credentials, they can use them to gain access to more and more systems and data within an organization. This is because many organizations use a single sign-on system, which means that once a user logs in with their credentials they can access multiple systems and services without needing to enter their credentials again.
- The lack of distributed/remote system recovery plans can create cybersecurity challenges due to dependency on a centralized system, business continuity risks, data loss risks, inability to respond to security incidents, and compliance requirements. Once a threat gets past network security, it cannot do anything to secure endpoints. Endpoints such as laptops, servers, and mobile devices are the entry points for attackers, and they can be easy targets to exploit compared to a secured network. Attackers can gain access to sensitive data in a matter of seconds, causing serious damage to an organization.
- External hackers breach patients' and medical systems originating from sources external to the health-care organization, aiming to pilfer and procure data, primarily driven by financial motives. One illustrative scenario involves the fraudulent submission of health insurance claims employing patients' personal data, thereby exemplifying how such breaches serve pecuniary interests. The Office of Inspector General contends that medical identity theft not only disrupts the continuum of patient care but also exacts a toll on public resources. Another

manifestation of external breach is exemplified by hackers who levy ransom demands on health-care institutions, seeking the restitution of patient data systems. According to findings by Verizon, the year 2020 witnessed a higher prevalence of external actors, as opposed to internal entities, compromising health-care data (51% vs 48%, respectively).

- Patient data theft for monetary gain or evil purpose is a common example of insider abuse. Other examples of insider abuse include convenience (overriding security rules to facilitate work) and curiosity (unapproved access to data not connected to the delivery of health-care). The residual occurrences of insider abuse encompass inadvertent actions, typified by human errors such as inaccurate input of data into EHRs or inadvertent engagement with phishing emails.

- Contemporary digital solutions possess the capacity to supplant antiquated systems, thereby enhancing health outcomes and enriching the experiences of patients, health-care practitioners, administrators, and IT personnel. Additionally, these solutions have the potential to mitigate vulnerabilities.

- Upgrading a system involves spending money on new technology purchases and paying technicians. If there is downtime, a health-care facility may have less ways to generate revenue.

10.5 An artificial intelligence–based cyber attack detection system in e-health

This section discusses a model involving cyber secure vulnerabilities and risks. Safeguarding the security of e-health telemonitoring systems is assuming heightened significance in light of the emergence of novel and evolving threats within this domain. In direct response to this concern, a Cyber Attack Detection System (CADS) model is introduced, adeptly harnessing AI methodologies to detect anomalies in real time, negating the necessity for a dedicated security analyst. The CADS model is meticulously devised not only to discern potential attacks but also to elucidate the nature of malicious activities, furnishing health-care practitioners with explanations and presenting them with data indicative of suspected attacks for insightful feedback. This proposal is further underscored by an in-depth exploration of the scenario wherein a remote patient health telemonitoring system is compromised by hacking. By embracing the CADS framework, health-care institutions stand to augment their proficiency in identifying and countering cyber threats, culminating in the preservation of patient data integrity and the overall safety and security of their telemonitoring infrastructures. Cyberattack detection refers to the task of recognizing unauthorized individuals using a computer system, including those with legitimate access but who may be misusing their privileges. The aim is to identify any attempts to utilize the system without permission or to exploit existing privileges. The CADS is a software solution that automates the task of identifying potential cyberattacks. Its central function encompasses continuous monitoring, detection, and subsequent response to any unauthorized activities originating from either internal organizational personnel or external cyber adversaries. Anomaly detection is a widely used approach in cyberattack detection, which assumes that any cyberattack will cause deviations from normal patterns. Health-care providers can use remote health-care communication to provide diagnostic services and monitoring to people living in smart communities, but the security of these systems is crucial because any security threat can cause serious problems, including false diagnoses or delayed interactions, which can compromise patients' privacy, health, and even lead to death. To manage security issues in healthcare systems, anomaly detection systems based on ML are essential for assuring security and reducing risks such as attacks using false data injection.

However, the task of identifying attacks within Internet of Medical Things (IoMT) systems is laden with complexities. These intricacies arise from the need to balance computing capabilities, memory allocation, battery efficiency, low latency, and network bandwidth—requisites that traditional centralized methods such as stand-alone cloud computing struggle to fulfill. In the prevalent cloud computing framework, data stemming from IoMT devices is transmitted to and from the cloud. However, this architecture fails to meet the demands of real-time emergency scenarios, such as instances necessitating swift medical intervention like fall detection or stroke prevention. The extended recovery time for data within cloud computing proves inadequate for expeditious response by medical professionals. The architecture of CADS is presented in Fig. 10.4.

The design of the proposed CADS architecture is shown in Fig. 10.4. This framework places a heightened emphasis on the security of data transmission originating from IoMT sensors. These data are routed through a network of three interconnected processing modules: Clinical Pathway Anomaly Detection (CPAD), User Interaction Engine, and Explainer. The User Interaction Engine, a pivotal component, encompasses three submodules: the User Interface, User Feedback, and

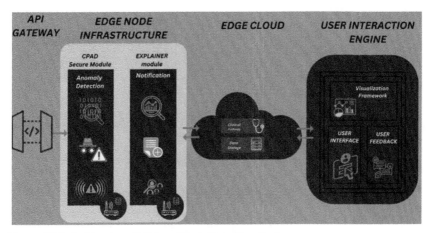

Figure 10.4 Architecture if Cyber Attack Detection System.

Visualization Framework mechanisms. Collectively, this engine oversees and orchestrates user engagements with the system. The CADS not only identifies anomalous data but also provides an explanation of why the data were classified as anomalous. This explanation is based on the explainable security (XSec) paradigm. If a home-care telemonitoring system is hacked, some electrocardiogram (ECG) instances may be classified as false positives or false negatives. After detecting these instances the user can interact with the data and analyze the reasons for the classification. This interaction is facilitated by the *User Interaction Engine*, which offers a dashboard for visual exploration of the data. Additionally, the *User Feedback* submodule facilitates an ongoing enhancement of classification accuracy and the efficacy of anomaly detection. This iterative process contributes to the cultivation of a more resilient threat identification capability. The overall system flow flowchart is denoted in Fig. 10.5.

In accordance with the architecture depicted, the technological framework is structured to facilitate patient care congruent with their specific clinical pathway (CP). This pathway encompasses a collection of diagnostic and therapeutic protocols pertinent to the patient's treatment regimen. The CP pertains to a mechanism devised for managing medical service quality by streamlining care processes to ensure optimal normalization. Its execution decreases the changeability in clinical practice and further develops results, targeting advancing coordinated and proficient patient consideration in view of proof-based medication, and to improve results in settings. A single CP can encompass multiple clinical guidelines across

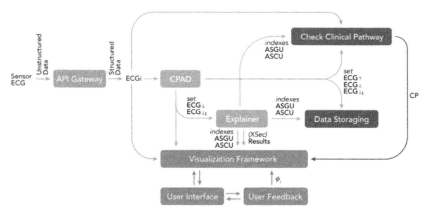

Figure 10.5 Data flow among system module.

various topics within a specific context. As a result, certain activities can be overseen by health-care professionals within medical facilities, while others can be self-managed by patients in a form of medically unsupervised fashion. The infrastructure for home care, denoted as IoMT-Edge-Computing, facilitates a distributed edge computing approach within the IoMT network, thereby minimizing latency and bolstering reliability. Consequently, patient monitoring devices are seamlessly integrated into the IoMT network. These edge devices then engage with a cloud infrastructure to amass clinical data and maintain communication with the respective medical personnel. This intricate framework does raise concerns regarding the potential vulnerabilities impacting the security of patient and clinical data.

The CADS, as outlined, integrates cybersecurity techniques to ascertain the specific data that have been compromised subsequent to a system breach. Additionally, it furnishes a comprehensive explication of the cyberattack event and facilitates interaction with identified irregularities that might manifest during the phases of remote and continuous patient monitoring and care. Central to this framework, the CPAD module undertakes the role of anomaly detection. Its formulation is presently tailored to align with medical service provisions within the health-care realm. The CPAD module plays a crucial role in averting the processing of incongruous or erroneous data, which could potentially endanger a patient's life. This objective is achieved through its capability to discern deviations from the patient's established CP. After the detection stage the Explainer module breaks down clinical information delegated irregularly and, through the Client Cooperation Motor, an approval demand is sent from a distance to the clinical staff. Uninterrupted telemonitoring of a patient's clinical metrics is conducted, obviating the necessity for a direct physical presence of health-care personnel. Varied clinical parameters, encompassing the blood oxygen level, electroencephalogram (EEG), ALT blood test, electromyography, ECG, and body temperature, are systematically acquired by the system via the utilization of IoMT sensors connected to the edge network, such as EEG. After that, clinical data of this kind are analyzed at the edge and a patient's CP is produced.

A patient diagnosed with congestive heart failure undergoes continuous monitoring at 15-minute intervals through a sophisticated terminal device designed to measure their ECG. Within the CPAD module the aggregated heartbeats are subjected to analysis to ascertain the presence of any abnormal measurements. Such anomalies could potentially indicate a breach within the telemonitoring system. The modular nature of the CADS facilitates integration with diverse IoMT devices through the utilization of Bluetooth technology. An API gateway assumes the role of validating the conformity of the acquired data. In the ongoing case study a sophisticated ECG monitoring terminal is interlinked with the CADS via IoMT. Input data undergo normalization through purpose-built APIs, facilitating seamless information interchange between the hardware devices and software components. This process simultaneously converts unstructured data into a structured format.

The proposed AI-based system aims to enhance the security of the telemonitoring infrastructure.

The system utilizes a CPAD module that analyzes structured data and detects any cyberattacks by retrieving anomalies. The CPAD module is based on Robust Deep Autoencoders (RDA), which are efficient in detecting cyberattacks by identifying and eliminating outliers and noise while discovering high-quality nonlinear features. Indeed, recent research endeavors have showcased the notable efficacy of deep learning methodologies in identifying cyberattacks. This is achieved through the application of neural architectures to execute anomaly detection, as highlighted in several studies [13−15]. As an illustration, should a cyberattack targeting ECG measurements be identified, a delayed response to the threat could potentially jeopardize the patient's care trajectory. Hence, the CPAD module leverages RDA for anomaly detection, ensuring swift responsiveness by promptly identifying aberrant ECG data. This module undertakes a straightforward binary classification of ECG data, classifying it into two categories: anomalous or nonanomalous. Typically, heartbeats are classified into five different types in medical literature [16], RonT—premature ventricular contraction, N—normal, UB—unclassified beat, SP—supraventricular, and PVC—premature ventricular contraction. To enhance the precision of the classification procedure, the CPAD module can first differentiate between normal heartbeats and anomalous ones. Subsequently, an additional RDA classifier is exclusively dedicated to identifying the four distinct categories of anomalous heartbeats. This systematic approach yields three distinct categories of classified ECG data: instances classified as normal, instances categorized as anomalous, and instances identified as belonging to anomalous heartbeats of specific types (PVC, UB, RonT, and SP). The output of the CPAD module results in the creation of a CSV file encompassing all instances that have been classified as anomalous and a confusion matrix for evaluating the performance of the classification model. The RDA method, based on the autoencoder approach, is used for anomaly detection by training the model on examples of anomalies and evaluating it on anomalous and normal datasets. The predicted instances are then broken down by the CPAD module into three distinct sets of heartbeat anomalies.

In our specific case study the User Feedback module plays a pivotal role in the system by facilitating the assessment of detected anomalies and incorporating feedback from medical professionals. This module is responsible for generating a feedback coefficient φ_i for each detection ECGi, signifying the doctor's feedback concerning a given instance. This module employs visual data mining (VDM) algorithms [17], which enable the application of diverse visualization techniques to effectively cluster data, thereby enhancing the data insight process through interactive means. This approach contributes to an improved understanding of the data landscape. The CADS can become stronger against external cyberattacks by using User Feedback from caregivers to confirm abnormal ECG detections. The User Interaction Engine uses VDM algorithms to group data and improve understanding. The User Interface allows caregivers to interact with ECG instances and improve classifier performance by indicating the correct class of anomaly [18]. The convergence of the CPAD, Explainer, and User Engine Interface modules will culminate in the presentation of anomalies in the form of threat insight on the user interface. This amalgamation will further entail the incorporation of a dashboard, which will serve as a resource for users to respond effectively to identified threats [19].

The security of telemonitoring systems is a significant challenge in the e-health domain, where threats are constantly changing. To address this issue a CADS has been proposed, utilizing AI techniques to automatically identify anomalies in the system. The system focuses on the task of detecting cyberattacks in a remote patient telemonitoring system and uses deep learning techniques to detect anomalous heartbeats. Cutting-edge XSec paradigms, complemented by novel explainable scores, contribute to the elucidation of detected anomalous heartbeats. These methodologies aid in comprehending the outcomes and facilitate their visual representation for the user. The system encourages user feedback, allowing for adjustments to the visualization reports and resulting in an enhanced dashboard. The implementation of AI algorithms yields a notable reduction in user reaction time, thereby rendering threat intelligence a pivotal instrument in bolstering system security and accomplishing automated threat analysis. In summation, the outlined system underscores the efficacy of AI techniques in augmenting user responses to threats within health-care telemonitoring systems.

10.6 Conclusion

In conclusion, AI-driven cybersecurity plays a vital role in protecting cyber-physical health-care systems from cyber threats and ensuring the confidentiality, integrity, and availability of critical patient data. Through the comparison of two case studies, "An Artificial Intelligence Cyber Attack Detection System to Improve Threat Reaction in e-Health" and "A Real-Time Healthcare Cyber Attack Detection Using Ensemble Classifier," it is clear that AI-based techniques are capable of quickly identifying and thwarting cyberattacks on health-care systems. The future of AI-driven cybersecurity in healthcare looks promising, with the potential to improve threat reaction and response times, reduce costs, and enhance the overall security of health-care systems. As the health-care industry becomes increasingly reliant on technology, it is critical to invest in AI-driven cybersecurity solutions to protect patient privacy and ensure the delivery of high-quality care.

References

[1] S. Mishra, S. Sahoo, B.K. Mishra, Addressing security issues and standards in Internet of Things, Emerging Trends and Applications in Cognitive Computing, IGI Global, 2019, pp. 224–257.

[2] A. Dutta, C. Misra, R.K. Barik, S. Mishra, Enhancing mist assisted cloud computing toward secure and scalable architecture for smart healthcare, International Conference on Advanced Communication and Computational Technology, Springer Nature Singapore, Singapore, 2019, pp. 1515−1526.

[3] M.A. Al-Shaher, R.T. Hameed, N. ȚăpuProtect, Healthcare system based on intelligent techniques, in: Proceedings of the Fourth International Conference on Control, Decision and Information Technologies (CoDIT), 2017, pp. 0421−0426.

[4] M. Marwan, A. Kartit, H.J.P.C.S. Ouahmane, Security enhancement in healthcare cloud using machine learning, Procedia Computer Science 127 (2018) 388−397.

[5] K. Kour, et al., An energy efficient routing algorithm for Wban, Turkish Journal of Computer and Mathematics Education (TURCOMAT) 12 (10) (2021) 7174−7180.

[6] Y. Qu, G. Zheng, H. Wu, B. Ji, H.J.S. Ma, An energy-efficient routing protocol for reliable data transmission in wireless body area networks, Sensors 19 (2019).

[7] A.A. AlZubi, M. Al-Maitah, A.J.S.C. Alarifi, Cyber-attack detection in healthcare using cyber-physical systems and machine learning techniques, Soft Computing 18 (2021).

[8] R.C. Aygun, A.G. Yavuz, Network anomaly detection with stochastically improved autoencoder based models. Available from: https://doi.org/10.1109/CSCloud.2017.39.

[9] P. Kumar, G.P. Gupta, R.J.C.C. Tripathi, An ensemble learning and fog-cloud architecture-driven cyber-attack detection framework for IoMT networks, Computer Communications 166 (2021) 110−124.

[10] I. Christov, G. Bortolan, Ranking of pattern recognition parameters for premature ventricular contractions classification by neural networks, Physiological Measurement 25 (2004) 1281.

[11] A. Hinneburg, D.A. Keim, M. Wawryniuk, Hd-eye: visual mining of high-dimensional data, IEEE Computer Graphics and Applications 19 (1999) 22−31.

[12] W. Meng, K.R. Choo, S. Furnell, A.V. Vasilakos, C.W. Probst, Towards Bayesian-based trust management for insider attacks in healthcare software-defined networks, IEEE Transactions on Network and Service Management 15 (2) (2018) 761−773.

[13] C. Ardito, T. Di Noia, E. Di Sciascio, D. Lofú, G. Mallardi, C. Pomo, et al., Towards a trustworthy patient home-care thanks to an edge-node infrastructure, in: International Conference on Human-Centered Software Engineering, Springer, 2020, pp. 181−189.

[14] C. Zhou, R.C. Paffenroth, Anomaly detection with robust deep autoencoders, in: Proceedings of the 23rd ACM SIGKDD International Conference on Knowledge Discovery and Data Mining, 2017, pp. 665−674.

[15] M. Sakurada, T. Yairi, Anomaly detection using autoencoders with nonlinear dimensionality reduction, in: Proceedings of the MLSDA 2014 Second Workshop on Machine Learning for Sensory Data Analysis, 2014, pp. 4−11.

[16] S. Chakraborty, K.S. Sahoo, S. Mishra, S.M. Islam, AI driven cough voice-based COVID detection framework using spectrographic imaging: an improved technology, in: 2022 IEEE Seventh International conference for Convergence in Technology (I2CT), 2022, pp. 1−7.

[17] L. Jena, N.K. Kamila, S. Mishra, Privacy preserving distributed data mining with evolutionary computing. in: Proceedings of the international conference on frontiers of intelligent computing: theory and applications (FICTA) 2013 (pp. 259−267). Springer International Publishing.

[18] T. Sivani, S. Mishra, Wearable devices: evolution and usage in remote patient monitoring system, Connected e-Health: Integrated IoT and Cloud Computing, Springer International Publishing, Cham, 2022, pp. 311−332.

[19] M. Kreuseler, T. Nocke, H. Schumann, A history mechanism for visual data mining. in: IEEE Symposium on Information Visualization, 2004, pp. 49−56.

Index

Printed and bound by CPI Group (UK) Ltd, Croydon, CR0 4YY

03/10/2024

01040427-0006